*

for small creatures

such as we

for small creatures such as we

Rituals for
Finding Meaning
in Our Unlikely World

Sasha Sagan

G. P. PUTNAM'S SONS
New York

PUTNAM
— EST. 1838 —

G. P. PUTNAM'S SONS
Publishers Since 1838
An imprint of Penguin Random House LLC
penguinrandomhouse.com

LIBRARY OF CONGRESS CATALOGING-IN-PUBLICATION DATA
Names: Sagan, Sasha (Alexandra Rachel Druyan), author.
Title: For small creatures such as we: rituals for finding meaning in our unlikely world / Sasha Sagan.
Description: New York : G. P. Putnam's Sons, 2019
Identifiers: LCCN 2019021246 | ISBN 9780735218772 (hardcover) | ISBN 9780735218789 (epub)
Subjects: LCSH: Sagan, Sasha (Alexandra Rachel Druyan) | Sagan, Sasha (Alexandra Rachel Druyan)—Family. | Rites and ceremonies—United States. | Spirituality—United States. | Children of celebrities—United States—Biography.
Classification: LCC GN473.S155 2019 | DDC 390.0973—dc23
LC record available at https://lccn.loc.gov/2019021246

International edition ISBN: 9780593087541

Printed in the United States of America
1   3   5   7   9   10   8   6   4   2

BOOK DESIGN BY KATY RIEGEL

*

For Helena Chaya,

light of my life

Contents

*

for small creatures
such as we

Introduction

I am a deeply religious nonbeliever. . . . This is a somewhat new kind of religion. —ALBERT EINSTEIN

A life without festivity is a long road without an inn. —DEMOCRITUS

When I was little and my dad was alive he would take me to see the dioramas at the American Museum of Natural History in Manhattan. This was a holy place for me, grand and full of answers to deep and ancient questions. It filled me with awe. But it also frightened me. I would hide behind my dad's legs, nervously stealing peeks at the frozen animals.

"That one just moved!" I would squeal.

"No, sweetie, it's your imagination. They can't move," my dad would tell me.

"How do you know?"

"Because they're dead."

This was a foggy and vague concept at the time. Foggier

even than it is now. I could see the bears and gazelles, but in some larger way they were not *really* there. I knew that my dad was deeply committed to accuracy. I knew he had more information on the topic than I did, so I could probably take him at his word; they were not moving. And yet my mind played tricks on me.

As we continued among the other exhibits—bright, sharp gems, early human and dinosaur skeletons, Neanderthal tools, gold Aztec figurines, Yoruba masquerade costumes— all sacred artifacts in the temple of history and nature, this mystery of death followed me. And I wondered about it a lot: what it meant, what exactly it was, what I was supposed to do with the knowledge of its existence.

"It's dangerous to believe things just because you want them to be true."

That's what my dad told me, very tenderly, not much later. I had asked him why I had never met his parents. He told me it was because they were dead.

"Will you ever see them again?" I asked.

The animals in the dioramas were dead, but we could still see them.

He told me that he'd like nothing better in the whole wide world than to see his parents again, but he had no evidence to prove he would. No matter how tempting a belief was, my father preferred to know what was true. Not true in his heart, not true to just him, not what rang true or felt true, but what was demonstrably, provably true. "We humans have a tendency to fool ourselves," he said. I thought about the dioramas again. He was right. The animals hadn't been moving, though I could have sworn they were.

We discussed the history of the world a lot at home. My parents taught me that there has never been any correlation between how true something was and how fervently it was believed. Sure, some things are subjective. This may be the best sandwich you have ever had or that may be the most handsome man you have ever seen. Those are in the eye of the beholder. For those phenomena that exist outside of our perceptions, things like the way the Earth moves, the causes of illnesses, or the distance between stars, there were objective realities. Maybe undiscovered by anyone on our planet, but no less real.

People once believed with all their hearts that the sun went around the Earth. But believing didn't make it so. There are certainly things we believe right now that will someday be revealed to be hilariously or abhorrently ignorant. Our understanding changes with new information. Or at least it ought to.

So if a person is interested in testing their preconceptions, in discovering how things really are and why, how does one go about that? My parents taught me that the scientific method is designed for precisely this job. My father was a scientist. He was the astronomer and educator Carl Sagan. Science wasn't just his occupation, it was the source of his worldview, his philosophy, his guiding principles. He and my mom, writer and producer Ann Druyan, taught me that belief requires evidence. They taught me that science wasn't just a set of facts to be compared and contrasted with other philosophies but a way of testing ideas to see which ones stand up to scrutiny. They taught me that what scientists think today might be disproven tomorrow, and that's wonderful, because that's the pathway to a better, deeper understanding.

This left me with something of a conundrum when my dad died. I had just turned fourteen. I longed to see him again somehow. I would often dream of being reunited with him. In these dreams, I would receive an elaborate explanation about where he'd been. Usually a misunderstanding or some kind of secret mission was to blame. The dreams would all end the same way. Elated, I'd tell him, "I knew it! I knew you weren't really dead! I have these dreams all the time where you come back—" And as I'd say those words, a sorrowful look would cross his face and I would realize. "This is one of those dreams, isn't it?" He would nod apologetically and I would awaken.

Still now, more than two decades later, I sometimes have those dreams.

Everything he taught me, everything he stood for, keeps me from believing that I will ever be reunited with him. But our secular home was not cynical. Being alive was presented to me as profoundly beautiful and staggeringly unlikely, a sacred miracle of random chance. My parents taught me that the universe is enormous and we humans are tiny beings who get to live on an out-of-the-way planet for the blink of an eye. And they taught me that, as they once wrote, "for small creatures such as we, the vastness is bearable only through love."

That is a line that appears in the novel *Contact*, the only work of fiction my dad ever published. He and my mom had first envisioned it as a movie, but movies take a very long time to make. This one in particular took about eighteen years, from conception to premiere. During those years, my parents decided to try the story out as a novel, which they dedicated to me. As with everything they wrote, *Contact* was a collaboration. Though that line gets attributed to my dad (his name is on the cover of the book), it's my mom who actually came up with those words, which have served as a perfect crystallization of our family philosophy.

Growing up in our home, there was no conflict between science and spirituality. My parents taught me that nature as revealed by science was a source of great, stirring pleasure. Logic, evidence, and proof did not detract from the feeling that something was transcendent—quite the opposite. It was the source of its magnificence. Through their books, essays, and television series, *Cosmos*, these ideas reached many millions of people around the world. So much of the philosophy contained in this book you're reading comes from them. My parents taught me that the provable, tangible, verifiable things were sacred, that sometimes the most astonishing ideas are clearly profound, but that when they get labeled as "facts," we lose sight of their beauty. It doesn't have to be this way. Science is the source of so much insight worthy of ecstatic celebration.

We humans are very good celebrators. I daresay we're the best at it in the whole wide world. Other species may express giddy enthusiasm when there's plenty to eat or when it's time to mate. Those events are certainly worthy. We celebrate them, too. But other species don't plan parties. They don't prepare feasts. They don't wear special outfits, put up special decorations, or say special words. They hardly even take the day off.

We, on the other hand, really get into it. Think of the billions of people who put time and effort into Christmas every year, into Chinese New Year, into Eid al-Fitr. Think of every culture, subculture, and religious sect that prescribes its own detailed procedures for honoring a holy day or a rite of passage. Think of every slight mutation as recipes and prayers are passed down from parent to child. Picture all the preparing, the decorating, the sacrificing, the riches spent over the eons to honor Zeus or Quetzalcoatl. In every inhabited corner of the Earth, human beings have created rituals to give order and meaning to existence. There is seemingly endless variety, and yet, there are clear and undeniable similarities, too.

I am the mother of a beautiful little girl, curious and lively, who regularly leaves me amazed. She is still too young to tell me her philosophy. I don't know what it will be or how she will come to it. As she grows and learns to think for herself, I can only hope to create a framework from which she will find that stirring beauty in our small place in the enormity. The more I thought about what I might be able to impart to her, the more I realized this framework might be useful to anyone who does not fit neatly into one system of belief or another, or into any at all. We all deserve holidays, celebrations, and traditions. We all need to mark time. We all need community. We all need to bid hello and goodbye to our loved ones. I do not believe that my lack of faith makes me immune to the desire to be part of the rhythm of life on this planet.

For me the biggest drawback to being secular is the lack of a shared culture. I can live without an afterlife, I can live without a god. But not without celebrations, not without community, not without ritual. There are no hymns about the testing of theories or mapping of genomes. No festivals to commemorate great inventions or medical break-throughs. Since I long for ways to honor the wonder of life, I've found myself making up new rituals. Sometimes I find I can repurpose the traditions of my ancestors to celebrate what I believe is sacred.

If you are devoutly religious, firstly, I'm delighted you're reading this. Thank you. If you have total conviction about your faith, you have plenty to celebrate already. This book is not intended to dissuade you, only to increase what there is to be joyful about.

If you are, like me, something else, maybe some combination of the words *secular, non-believing, agnostic, atheist,* or possibly *pagan,* my hope is that this book might help separate skepticism from pessimism. I don't think that faith is a requirement to see a world full of provable miracles and profound meaning. I also don't think lack of faith means you must give up your most beloved rituals. There is a way to honor your traditions and your ancestors without feeling you are just going through the motions.

I myself am only a few generations removed from some very religious people. My mother's grandparents were Orthodox Jews. And I mean orthodox. My great-grandmother Tillie kept a kosher house, and for her, that meant that if a dairy fork touched a piece of meat—by accident, for a second—it had to be buried in the backyard for a year. Her husband, my great-grandfather Benjamin, volunteered as the night watchman at their temple because they were too poor to pay the annual membership dues.

Tillie and Benjamin were born at the end of the nineteenth century in shtetls in parts of the Russian Empire that are now Latvia and Belarus. They got out while the getting was still relatively good, despite being virtually penniless. The first leg of their journey was to join Tillie's sister, who had already emigrated to Stockholm. The story passed down to me was that on their travels westward, every day Tillie and Benjamin would get a day-old loaf of bread from a bakery because they couldn't afford a fresh one. Then they would sit on the curb and say to each other in Yiddish, "No, you eat, I'm not hungry." Even though they were both borderline starving, neither would cave. My mother told me this when I was small. I have pictured that exchange tens of thousands of times over the course of my life, filling in details, letting it shape my idea of what true love is.

My great-aunts were born in Sweden, before the family traveled by steerage across the Atlantic to Ellis Island. My grandpa Harry was born in the New World in 1917.

Twenty years later, Harry was a journalism student at New York University, the first person in his family to get anywhere near college. His parents might have thick Yiddish accents, and be unable to read or write in English, but he was American, and studying in Manhattan was making him cosmopolitan and skeptical.

One day he rode the train home to Queens, working up the nerve to talk to his father about something important. At home he found him davening, wrapped in a tallis, lost in prayer. When Benjamin opened his eyes he was delighted to see his only son, his college boy, standing before him.

Harry told his father he would no longer keep kosher, no longer pray, no longer spend Friday nights at shul. Because he just didn't believe. Not in the teachings he was brought up with, not in the Torah, not even in God.

He braced for his father's reaction.

I've often imagined the weight of this moment, too. The guilt of knowing what your parents sacrificed to escape oppression, how hard they worked to preserve their way of life, how carefully they taught their beliefs to their children. And knowing that across the ocean in their homelands, at that very moment, the political climate was turning, and your people were starting to disappear.

But, safe in New York, my great-grandfather looked up and smiled at his son and said the immortal words: "The only sin would be to pretend."

Decades later, by the time this story was passed down to me in vivid color through my mother's impeccable storytelling, those words had become a kind of family mantra. Even though that marked the end of possibly thousands of years of devout Jewish belief, it reaffirmed a different element of Jewish tradition: debate, philosophical questioning, skepticism.

It wasn't like my grandfather all of a sudden wasn't Jewish because he renounced the beliefs of his ancestors. Judaism in particular has a funny way of blurring the lines between religion, culture, and ethnicity. For example, when I take one of those DNA test kits, I get a result that says I am an Ashkenazi Jew. I don't think there's a test that can tell you if you're, say, a Presbyterian. At least not one that uses your saliva. And it's not just a matter of what a genetic testing service thinks. I see myself as a Jew even as I sit here writing a book about my lack of faith. It's complicated. It took my husband, Jon—whose parents were raised Protestant and Catholic respectively but is neither himself—a while to get it when we first started dating. We'd have conversations like this:

HIM: "You're Jewish, but you don't believe in God or anything religious?"

ME: "Yes."

HIM: "So you're atheist or agnostic or something?"

ME: "Yes."

HIM: "So not Jewish?"

ME: "No, still Jewish, too."

This went on for a while. But eventually he came to understand that even though I don't subscribe to the supernatural elements, I am still a Jew.

HIM, NOW: "It's an ethnic group, a culture and a religion, a Venn diagram that overlaps a lot but not completely. Just as there are Jewish converts who are not ethnically Jewish, there are ethnic Jews who are not religious."
ME, NOW: "Yes!"

But this is not a singularly Jewish issue. Lots of people don't, for instance, consider Jesus Christ their Lord and Savior but still take pleasure in Christmas. It doesn't even have to be about religious identity. How many Americans of, let's say, Italian ancestry identify as Italian but speak only a few words of the language here and there, have read no Dante, seen no Botticellis, never even stepped foot on that boot-shaped peninsula? To a Florentine this person is not their countryman, but back in the Bay Ridge section of Brooklyn, he sees himself as Italian.

Elsewhere in Brooklyn, in the still-ungentrified parts of Williamsburg, most Orthodox Jews would not see me as one of them, should they happen upon me some Friday night, bare-legged, sharing a dozen oysters with my Gentile husband. But when I see them, I know we are connected. I know that in some alternate universe, maybe out there in the multiverse, there is a world where my grandfather never gets up the courage to admit his nonbelief and goes along with his parents' traditions. There's some version of me who speaks Yiddish at home, keeps a kosher kitchen, dresses modestly, and genuinely believes. Or at least doesn't feel she can tell anyone that she doesn't.

But that's not what happened. My granddad said his piece and raised my mom and her brother as secular Jews, and she went on to do the same. Never letting go of the Jewish part completely.

My great-grandparents' beliefs gave shape to everything they did. I have different convictions than they, but I envy the way meaning infused their lives.

Through my secular lens, I see a different meaning in their traditions. In a way, it's really science that's been inspiring rituals all along. Beneath the specifics of all our beliefs, sacred texts, origin stories, and dogmas, we humans have been celebrating the same two things since the dawn of time: astronomy and biology. The changing of the seasons, the long summer days, the harvest, the endless winter nights, and the blossoming spring are all by-products of how the Earth orbits the sun. The phases of the moon, which have dictated the timing of rituals since the dawn of civilization, are the result of how the moon orbits us. Birth, puberty, reproduction, and death are the biological processes of being human. Throughout the history of our species, these have been the miracles, for lack of a better word, that have given us meaning. They are the real, tangible events upon which countless celebrations have been built, mirroring one another even among societies who had no contact.

As I see it, here we are on this rock that orbits a star, in a quiet part of a spiral galaxy somewhere in the great, wide vastness of space and time. On our rock these events, changes, and patterns have an enormous impact on us Earthlings. They are important to us. We have spent a lot of time trying to decode them, to manage our expectations, to predict what's coming, to grow, to thrive, to survive. No matter when or where on Earth we live, we humans tend to schedule our most important events around the same times. Sure, Christmas and Hanukkah often fall around the same week. But so does the Dongzhi Festival in China, Umkhosi Wokweshwama among the Zulu, Yaldā in Iran, and Soyal among the Hopi of the American West.

And it's not just certain times of year. It's times of life, too. Every culture from the Amish to the Maasai has coming-of-age rituals that, at their core, are the same as any bar mitzvah, *quinceañera*, or sweet sixteen you've ever been to. Not to mention the vast array of human ways to welcome a newborn, marry a couple, or honor the dead. Ecstatic joy to deepest sorrow, the heart of these rituals lies beyond belief.

While our calendars have shifted, and our climates, politics, and superstitions vary, somewhere in the depths of whatever you celebrate there is very likely a kernel of some natural occurrence. We needn't resort to myth to get that spine-chilling thrill of being part of something grander than ourselves. Our vast universe provides us with enough profound and beautiful truths to live a spiritually fulfilling life.

Nature is full of patterns and we humans love finding them, creating them, repeating them. That's at the core of language, math, music, and even ritual, which is the repetition of words or actions deemed worthy of representing something bigger than ourselves. Some rituals are very private, some are very public. Some are so commonplace we don't even think of them as rituals. My view is that all over the world and across time, these are all a form of art, an elaborate performance or a secret poem, all vital in their ability to help us face the nature of time and change, life and death, and everything else we cannot control.

So much of human culture is designed to help us come to terms with the most astonishing elements of existence. Every single one of us appears seemingly from nowhere and then, eventually, returns to nowhere. We are conceived, we grow, and we die, but what happens beyond that is a great, haunting mystery. We grapple with it by marking how and when things change here on Earth, both cyclically and permanently.

I believe rich, meaningful rituals can be modern. You

can invent one tomorrow. "It's just the way things have always been" has all too often been used to exclude and demean, or to justify the odious. An old tradition is not intrinsically better than a new one. Especially when it is such a joy to make new ones up—ones that reflect exactly what you believe, ones that make sense of your life as you experience it, ones that bring the world a little closer to the way you wish it could be.

It's not always easy to start something from scratch, though. It can feel a little contrived, a little ridiculous. You can lose that sense of inclusion in a community, being part of a legacy. Without going full Tevye, there is something deeply reassuring about performing the specific steps, the exact motions, that your grandparents performed, and that they learned from their grandparents. You say a prayer, light a candle, make a dessert just the way they did, and imagine the nameless generations stretching back into the past. There is a pleasure in this. It's a kind of connectedness, a kind of time travel, providing a sense of certainty in what's tried and true. That's why new cultures are so often built upon existing cultures. Elements are borrowed, repurposed, reinterpreted, appropriated, stolen, or used to quell unwilling converts. Some new secular scientific tradition would undoubtedly borrow from theistic ones. For example, throughout this book I have found it impossible not to use the language of belief. Words like *sacred*, *magical*, and *spiritual* come from theism, but they describe the same feeling even when it's elicited by an understanding of scientific phenomena. These words are evolving with our understanding of our place in the splendor of existence.

For years, people told me that once I had children I would become more religious. These weren't fervently religious people necessarily; in most cases they were not even believers, just people who felt that children need traditions and that that's what religion provides. I would usually get a little defensive. But they were partially right.

I do now have an increased urge to celebrate things with our daughter. I love and have always loved special occasions, the break from monotony, a reason to put on a dress, the banter, feasts, merriment, and the feeling of being part of a group. I love parties, the marking of time, the sensation of feeling it pass. And I want to provide that for my child. I want to make her face light up. I want to evoke that gleeful giggle, that sense of wonder that makes life on Earth feel so magical, so intentional, when you're little. But I can't go through the motions. I can't bring myself to tell her anything I don't believe is true.

So I find myself eager to map out a year that is sometimes inspired and informed by the practices and beliefs of her ancestors on both sides, but not shackled by them. I want to create moments that make us feel united with other Earthlings, without the dogma that divides us. Religion, at its best, facilitates empathy, gratitude, and awe. Science, at its best, reveals true grandeur beyond our wildest dreams. My hope is that I can merge these into some new thing that will serve my daughter, my family, and you, dear reader, as we navigate—and celebrate—the mysterious beauty and terror of being alive in our universe.

chapter one

Birth

Yesterday a drop of semen, tomorrow a handful of [...] ashes.
—MARCUS AURELIUS

After our daughter was born, Jon and I said to each other a thousand times a day, "I can't believe she's here!" "I can't believe we have a kid!" "I can't believe we made a person!" Every day for months and months we said it out loud as if we were just discovering how reproduction worked. We struggled to wrap our minds around it. I actually don't suppose I'll ever truly get over this idea. My mother never has. She sometimes still joyfully says to my brother Sam and me, "You don't understand, you didn't exist, and then we made you! And now you're here!" We roll our eyes and say, "Yes, Mom, that's how it works." Which is true, but no less astonishing, beautiful, or thrilling. Being born at all is amazing. It's easy to lose sight of this. But when a baby comes into the world, when a new human appears from inside of another, in the accompanying rush of emotion, we experience a little bit of the immense brazen beauty of life.

Rituals are, among other things, tools that help us process change. There is so much change in this universe. So many entrances and exits, and ways to mark them, each one astonishing in its own way. Even if we don't see birth or life as a miracle in the theological sense, it's still breathtakingly worthy of celebration.

Typing these words, I am, like you, experiencing the brief moment between birth and death. It's brief compared to what's on either side. For all we know, there was, arguably, an infinite amount of time before you or I was born. Our current understanding is that the big bang gave birth to the universe as we know it about 13.8 billion years ago. But the big bang may or may not be the beginning of everything. What came before, if anything, remains an unsolved mystery to our species. As we humans learn, create better technology, and produce more brilliant people, we might discover that which we currently think happened is wrong. But somehow, something started us off a very long time ago.

In the other direction there will, theoretically, be an infinite amount of time after we're dead. Not infinite for our planet or our species, but maybe for the universe. Maybe not. We don't know much about what that will entail except that the star we orbit will eventually burn out. Between those two enormous mysteries, if we're lucky, we get eighty or one hundred years. The blink of an eye, really, in the grand scheme of things. And yet here we are. Right now.

It's easy to forget how amazing this is. Days and weeks go by and the regularity of existing eclipses the miraculousness of it. But there are certain moments when we manage to be viscerally aware of being alive. Sometimes those are very scary moments, like narrowly avoiding a car accident. Sometimes they are beautiful, like holding your newborn in your arms. And then there are the quiet moments in between, when all the joy and sorrow seem profound only to you.

On one particular day a few winters ago I felt this intensely. I had just found out that I was pregnant, full of wonder and nausea. Everything was about to change forever. It was also the twentieth anniversary of my father's death. Twenty years feels like a shockingly long time. It's significantly longer than the time I had with him. I miss him very much. Sometimes, still now, so much that it feels intolerable.

Feeling the entrance of one new being and the loss of another brought on a series of paradoxical emotions, and a powerful sense of my place in the universe. I remember walking around the city, stunned that everyone I saw, the owner of every wise and wizened face, was once a baby. This seemed revelatory, despite its obviousness. I couldn't help reflecting on how any of us got here in the first place. Human beings do not go back to the beginning of this universe. In our present configuration we've only been around about a few hundred thousand years—the number changes as we uncover more of our fossilized ancestors—but the planet we live on is more than 4.5 billion years old. We're new here. We evolved from slightly different creatures who evolved from somebody else and so on back to one-cell organisms that we would not recognize as our relatives, but nonetheless, they are. How those one-celled forebears came to be is just now beginning to become clear. Even less clear is how exactly it will end for us: we will either destroy ourselves, be destroyed by an outside event, or evolve into something unrecognizable.

As the small creature inside me expanded my midsection, I was reminded of how many pregnant girlfriends over the years have looked at me with a kind of mild, jokey horror and exclaimed, "It's like there's an alien inside me!"

My dad spent a lot of time thinking about aliens, trying to determine if they existed. He never found out, because so far there's no evidence we've ever had contact with life from elsewhere in the universe. For my dad, as for me, belief required evidence. To say "I don't believe" in something doesn't mean that I am certain it doesn't exist. Just that I have seen no proof that it does, so I am withholding belief. That's how I think about a lot of elements of religion, like God or an afterlife. And it's the same way my dad thought about aliens. As he once said, "Absence of evidence is not evidence of absence." We don't have proof, so we don't know. And yet we all seem to have a vivid idea of what an alien is like. We almost always imagine they look like us but they're smaller. They have large eyes and no hair. They don't talk. They don't know the social mores. They might be good or they might be evil, but they definitely want something from us and as soon as they arrive, everything will be different forever.

Babies are not like aliens. Our idea of aliens is like our idea of babies.

Maybe that's part of what my dad was thinking when I arrived. My mother tells me that when I was born, my father lifted me up, looked at me, and said, "Welcome to the planet Earth."

Then they didn't name me for three days.

When they finally did, I got the middle name Rachel, for my dad's mom. She was both magnetic and impossible, a mesmerizing storyteller with a one-of-a-kind laugh. She had a very difficult childhood. Her mother died in childbirth when Rachel was two. Her father (who may or may not have come to America to escape a murder rap in Russia) sent her back to Europe to live with aunts she had never met until he remarried a few years later. But Rachel grew up in New York, found true love with my grandfather Sam, and in many ways made my father who he was. It's a complicated legacy.

When I was a small girl, family members were often astonished, alarmed even, at how clearly my mannerisms resembled hers. It was not learned behavior. I was born close to nine months to the day after her death. My parents would get chills at the sound of Rachel's distinctive laugh emerging from their little daughter. It was "very eerie," I was told. It would have been easy for me to make a leap from these reactions to something ominous, something scary. I might have guessed that I was possessed by my dead grandmother, or that she was somehow haunting me.

When I was eight, my younger brother was born, and named for our grandfather Sam. Soon he bore such a resemblance to our father that, when invitations to my dad's birthday party went out with a black-and-white picture of him as a little boy swimming off Coney Island, people called to say, "Yes, we can come to the party, but why is there a picture of Sam on the invitation?" To my parents these family resemblances were something wondrous.

My parents told me that there was a kind of secret code called DNA running through our veins. I learned it carried the traits of ancestors I would never meet. My genes linked me back to the earliest humans, to prehistoric mammals and back eventually to the first life on Earth. And if, someday, I had children of my own, I would become a link in the chain, passing along an embedded part of myself to the future generations who would never know my name. This was, to me, more satisfying than any other possible explanation. And it was verifiable, independent of my belief or lack thereof.

This was my introduction to a world of giddy enthusiasm about the fact that the universe is bigger than we are currently able to comprehend, that we live on a planet we are perfectly adapted for, that we are capable of critical thought, and that our understanding of all this grows deeper and more astonishing with time. And that, as far as we can tell, this all happened by chance. Think of the asteroid that could have just missed the Earth, sparing the dinosaurs, robbing those little Cretaceous mammals of the chance to flourish and eventually evolve into you and me. I find it impossible not to think of this as miraculous, despite the connotations.

Even with our species flourishing, the chances of any one of us being born are still remote. Think of all the slight variations in human migration patterns, for example, that could have kept your great-great-grandparents from ever crossing paths. If you have any European ancestry, someone in your lineage had to survive the black death in the fourteenth century, which killed more than half the people on the continent. If you have any Native American heritage, somehow your forebears managed to pass their genes on to you, despite the fact that only 10 or 20 percent survived the microbes and violence brought by European invaders. Whatever your ancestry, the list of wars, raids, plagues, famines, and droughts your genetic material had to overcome is stunning. All this in order to arrive at the moment where you, exactly you, are ready to depart your mother's womb and come into the great wide world.

Let's say there were three decisive moments in each of your biological parents' lives that led to their meeting. This is a ridiculously conservative estimate; it's probably millions of moments, but, for simplicity's sake, let's say three. Your mother chose to go to such-and-such university, she chose to strike up a friendship with so-and-so, and years later, she chose to accept so-and-so's invitation to the party where she met your dad. Meanwhile, your dad chose X career, where he met X colleague, and eventually accepted the invitation to the party where he meets your mom.

At the risk of stating the obvious, in order for your parents to meet, they each had to be born, which required both sets of *their* parents to meet. And before that, your grandparents had to be born, so your great-grandparents had to meet. And so on and so on, all the way back to the first humans in East Africa.

Right now we think there have been approximately 7,500 generations of *Homo sapiens*. They all had to find each other in that perfect moment. There are so many forks in the road that within ten or fifteen generations the odds become mind-boggling.

But we've only accounted for conscious decisions. What about happenstance?

My mother's parents met on the New York City subway. In a car of the E train during rush hour. It was 1938. My granddad Harry was reading William Faulkner's *Absalom, Absalom!* and when he went to turn the page my grandmother Pearl put her hand on his and said she wasn't finished reading. How many different cars were on that train? How many different trains came through that station? How many different train lines could they each have lived on? How many different cities could their parents have emigrated to? And so on and so on back for all of history.

It's true that the vast majority of unions that led to any of us were not rom-com-worthy meet-cutes. Many were terrifying wedding nights, a stranger at the right time or place, a warm body in the cold, lonely darkness, and unspeakable horrors at the hands of invaders and enslavers. But there were, undoubtedly, also some unions formed in the glorious rapture of true love.

For every single woman who led to you, there would have been a moment of clarity. Her period wouldn't come. Instead maybe nausea and sore breasts. Soon, her body would start to change, her growing belly a badge of pride or shame. Sometimes I imagine all the inner thoughts of these women, their excitement and fear. Thousands of stories of passion and pain, ecstasy and agony, all but a few lost to the ages, but all equally critical to you being born, to you being genetically who you are.

On the other hand, if you believe in destiny or determinism, you believe that only one event had to happen: the universe had to start. Everything after that is, was, and will be inescapable. Including you reading this sentence. Or you might believe some events are inescapable and others are not. The ideas that "everything happens for a reason" or that certain things are "meant to be" are often offered as reassurances. But, to me, they are not as astounding or awe-inspiring as the idea that, in all this chaos, somehow you are you.

You do not need to formulate an opinion on the nature of free will and fate to know being born is profoundly special. The arrival of a baby is a cause for celebration all over the world: baptisms, baby namings, circumcisions, ear piercings, or other kinds of ritual scarification are popular to mark the occasion. Umbilical cords and placentas get incorporated into an array of traditions. Sometimes the welcoming ritual is a tiny private act. For example, among Hindus and Muslims there is a belief that a baby's first taste should be sweet—a drop of honey or a piece of fruit is used to introduce new taste buds to the world—and that a sacred prayer should be the first thing they hear. Sometimes it's a feast once you've settled into life on Earth, as it was for

ancient Incan babies upon being weaned and for generations of Chinese babies when they reach one hundred days
of life, a great accomplishment during the eons when infant
mortality rates were high. Now it's a reminder of how entwined birth and death can be.

In Côte d'Ivoire, the afterworld of the Beng people is a
large city where the dead speak all the languages of the
world. When a baby is born, reincarnated into our world,
they are sometimes given colonial French coins, the currency of their netherworld and of the power that upended
their society here on Earth. But the first gift a Beng baby
gets is a cowry shell, the currency of their ancestors before
the French arrived, another kind of token of remembrance
for their previous life.

Every ritual, tradition, superstition, and celebration designed to welcome a baby has, at its heart, a hopeful wish for
that new human.

Jon and I had all the hopes and dreams that come with
new parenthood, but we didn't have an ancient framework
to express it. We didn't have a symbolic gift to offer our
daughter, no special timeworn words to whisper in her
teeny ear, except "Welcome to the planet Earth"—a tradition only one generation old. Maybe this is because we
didn't have a religious community that expected this of us.
Maybe it's just because we were too tired and overwhelmed
to imagine entertaining. If we had marked her arrival with

a large group of friends and family, with ritual and tradition, we might have had an easier time processing this enormous change. Maybe we wouldn't have spent her first year in shock that we had created life. Or maybe we would have increased our sense of awe by celebrating it.

In retrospect it seems so obvious that our daughter deserved to be welcomed in some formal way. Maybe with a party, or some secular ceremony we could have invented. Maybe the reading aloud of a passage or poem that felt fitting. But devising rituals takes a little vision and creativity, and I had spent all of mine on preparing her room, picking out her tiny clothes, and imagining what kind of mother I would aspire to be. Maybe someday we'll have another baby and be better prepared with a ritual for the occasion.

I don't know exactly what that might be, but there is one ritual I wish we'd at least explored. It's beautiful, tangible, and appears among disparate cultures: the planting of a tree. A tiny seed is deposited in Mother Earth and soon a new life begins to emerge.

In the Balkans it's a quince tree. In parts of China it's an empress tree—specifically when it's a girl. In Jamaica, the baby's umbilical cord is sometimes buried alongside the seed. There is a Jewish tradition, with Talmudic roots, that calls for the planting of cypress trees for girl babies and cedar trees for boys. When they grow up, their chuppah (the

Jewish wedding canopy) is built from these branches. In Piplantri, Rajasthan, in northwest India, the villagers plant 111 trees when a little girl is born: mango, North Indian rosewood, Indian gooseberry. In a world where the birth of a daughter was often seen as a burden, this ritual celebrated it, despite being rooted in grief. It was the invention of a man who once served as Piplantri's *sarpanch* (a position akin to small-town mayor). His beloved daughter died very young and he wanted to honor her memory. The villagers planted many thousands of trees but worried termites would destroy them. So the villagers started planting aloe vera, which they believed would kill the pests. Soon there were millions of aloe plants in Piplantri. They didn't know what to do with them all. But the women of the village learned to make juice and salve, and soon the village grew prosperous, an unexpected blessing born out of their ritual.

The Jewish, Chinese, and Balkan variations of this tradition are very old. This Rajasthani version sounds ancient and maybe apocryphal. It's not. It started in 2006. (There is another element to this tradition that I love. When the baby girl is born and the trees are planted, the girl's parents sign a legally binding affidavit promising that their daughter will not be married off before she is of age, that they will support her in her education, and that, together, they will care for these hallowed trees as they, like their daughters, grow skyward.)

For the people of Piplantri, the planting of 111 trees is meaningful. They don't need an ancient proclamation or divine vision to celebrate birth.

As my own daughter grows skyward, I find myself looking into her eyes and glimpsing our ancestors, the ones we have known and the ones we never will. When I look into her little face I can't imagine her being anyone else. But I know that somewhere along the road that led to her existence, something could have unfolded very differently. I feel a sense of awe for every single thing that happened to bring us to this exact moment, where we are each *us*, alive, experiencing life together on this world.

*

chapter two

A Weekly Ritual

*Religion isn't about believing things. . . . It's about behaving in
a way that changes you, that gives you intimations of holiness
and sacredness.* —KAREN ARMSTRONG

R eligions don't agree on which day is holy, but by and
large they do agree that once a week you must check in
with your beliefs, your community, and yourself. For Jews,
this ritual starts Friday at sundown and lasts until sundown on Saturday. For Muslims, it usually starts with Friday afternoon prayers. For the wide range of denominations
of practicing Christians, it's on Sunday, with the exception
of Seventh-day Adventists, who are defined by their obser-
vance on Saturday. For Quakers, a sect of Protestants who
downplay annual holidays because they see every day as an
equal celebration of Christ, silent worship at their weekly
meetings is the heartbeat of belief. In Buddhism, the holy
day of the week changes with the phases of the moon.

For my great-grandparents the holy day was called Shabbos (the Yiddish word for the Sabbath; Shabbat is Hebrew). Each week, for an entire rotation of the Earth, they did not work or handle money or use electricity: no lamps, no phones, and no riding in cars. Just prayer, synagogue, and family time. It was about rest, reflection, and taking stock of the week, a kind of early TGIF.

There are, as with all things, loopholes even for the orthodox. For example, there are large buildings in places densely populated by observant Jews where the elevators are preprogrammed to stop at every floor during Shabbat. If you're not pressing any buttons, you're not breaking the rules, right? This is the subject of much debate. Other exceptions also get made. If you're extremely ill, it might be okay to ride to synagogue as long as a Gentile drives you. But even when my great-grandfather Benjamin was dying of stomach cancer, he was so devout he still walked to services every week.

After Grandpa Harry's declaration of independence, holidays like Passover and Hanukkah remained part of the family calendar, but we lost Shabbos, and some part of me has longed for it.

Luckily, secular people have weekly rituals, too. Maybe Thursday happy hour with coworkers, a steady date night, a favorite exercise class, or a time set aside to volunteer somewhere. Whether it's a trip to the farmers market, the nail salon, or the psychiatrist's office, almost everyone I know has something particular they do once a week. Before VCRs, DVRs, and streaming services, a favorite television show, broadcast at its appointed time, was a ritual. Sporting events are still watched mostly as they unfold live, millions of people sharing thrills and agonies in their living rooms. These may not leave us with a sense of the divine, but they create a pattern for our lives, a set moment to dip back into our communities and ourselves.

After school on Fridays my mom would usually take me to buy a loaf of challah from the Clever Hans Bakery in Ithaca, in upstate New York, where I grew up. This was a kind of hat tip to the ways of our forebears. For me, the visit to the bakery was more about the brownie with mint icing that I called a "greenie," and that I ate ferociously in my car seat on the way home. This was another kind of holy sacrament, my first appreciation of what it felt like to finish the week, to transition from work to rest (even if, for me, at that time, work largely consisted of coloring). It was, I think now, a way to learn about marking time, about feeling it happen. Time is an elusive concept. It's passing constantly, yet it's so hard to *feel*. It's like lying in the grass, trying to feel the Earth rotate. When changes are both small and constant, we can't grasp them. But watching a sunset, for example, we can process that we've successfully completed another rotation.

On most weekends my mother and I had another sacred

ritual, although I have only started thinking of it as one from the purview of adulthood. My mother would produce a large book of construction paper, some Elmer's glue, a pair of safety scissors, and one large piece of cardboard. We would decide on a theme—the ocean, space, dinosaurs, forests— and create a world by cutting out paper flora, fauna, rocks, and suns. Sometimes the themes were even religious. The stories of the Garden of Eden and Noah's Ark were not censored in our house. They were taught. I cherished my wooden toy ark with two of every animal and displayed it prominently. The only difference was that these pillars of civilization were presented as important, influential literature—not history. (My mother would say things like "There is evidence that there was a great flood, but the fossil record contradicts the idea that two of every animal survived and repopulated the planet.")

It didn't really matter what shapes we cut out, it was more about the time we spent together. Sitting and talking, doing something a little educational and creative. It was about the ritual. And, I think, the weekly timing. Every day would be too much, not special enough, and too time-consuming. Once a month would be too infrequent; we'd lose the rhythm of it. There's something about once a week that's just right.

The week doesn't exactly have an innate astronomical root the way the seasons do. It may be linked to the phases of the moon, which are about twenty-eight days and evenly divide into seven nicely. Many cultures appear to have adopted this unit independently of one another.

But a week doesn't *have* to be seven days. Various ancient cultures had eight-, nine-, and twenty-day weeks. During the French Revolution, when everything else in French society was being questioned, the calendar was briefly revamped into a base-ten system. The ancient Egyptians had that same idea.

The way we experience time has so often been entwined with our beliefs. This is evident in another part of our weekly calendar. In English, *Monday* is a contraction of *moon* and *day*. In Spanish, French, and Italian, days share the same basic construction. It might not sound like it in English, but Tuesday is for Mars. The romance languages make it obvious: *martes* (Spanish), *mardi* (French), and *martedi* (Italian). Wednesday, also known as *miércoles, mercredi*, and *mercoledí*, is for Mercury. Thursday is for Jupiter, Friday for Venus, Saturday for Saturn.

Sunday you can work out for yourself.

All these names of celestial bodies were also the names of the gods in ancient Europe. (For this reason Quakers sometimes refer to the days of the week by number instead of their names, so as not to pay homage to pre-Christian idols.) From Greece to Scandinavia, people imagined their gods were the personifications of the planets and stars, that the workings of the universe and the gods were one system. In fact, all over the world, for most of history, nature and religion were not just intertwined but inextricable. The universe was sacred. The gods and nature were not yet at war.

I spent a lot of time reading ancient myths as a child. You know the kind—the ones where a god, disguised as an unassuming human shepherd or the like, comes down from Mount Olympus to meddle in the earthly affairs of hapless mortals. And once, in adulthood, I had the sense that I was living just this kind of mythical trope. It was during one of the most impactful chance encounters of my life.

Jon and I were just married when we found out we had to be in Washington, DC, the week of my birthday. The Library of Congress was celebrating the acquisition of my parents' papers. It was a huge, emotionally draining event. I wept uncontrollably during the speeches and videos, seeing my dad young and healthy, hearing his voice.

Jon was determined the event wouldn't completely swallow up my birthday. He somehow managed to secure a reservation at a restaurant we'd been fantasizing about since Anderson Cooper's awestruck report on it months earlier on *60 Minutes*. It was the kind of place where the food wasn't even really food so much as gastronomic art—mojito orbs, popcorn in liquid nitrogen. We knew the meal would be surreal. We didn't know the cab ride to the restaurant would be, too. We were discussing Jon's career as we hailed the taxi. He was in a rut, feeling unchallenged, uninspired. I told him, not for the first time, that I thought he should pursue something he loves; that if he wasn't happy, maybe he should go to grad school, change careers. He reminded me, also not for the first time, that he had a good, stable job, and leaving it for the unknown would be irresponsible. We hopped in, gave the driver the address, and continued the familiar debate—already well versed in the grand marital tradition of having the same conversation again and again after living together for more than six years.

"How long have you been married?" the driver asked in a melodic accent, looking at us in the rearview. He barely glanced at the road.

"Stoplight coming up, sir," Jon said.

"Six weeks," I said.

"Six months—you're just married!"

"No, six *weeks*!"

The driver let out a low, vowely sound, conveying a wide breadth of knowledge on the topic of marriage. Then he began to sing: *"Oh Lord, won't you buy me a Mercedes-Benz . . ."*

Jon and I looked at each other. The driver's song morphed into a medley of other Janis Joplin numbers, then into a series of ballads in a language we didn't know. He serenaded us, occasionally looking back at the road. Jon put his arm across me as an extra seat belt.

"Do you know Janis Joplin?" the driver asked.

Yes, of course, we said.

"That is what I like about America. In Sierra Leone, each village has its own songs. Here, everybody knows the same ones. Do you sing?"

I'm nasal and rhythmless. Jon less so, but no, we said, we don't sing.

"You must sing together!" The driver was emphatic, as though our marriage was lacking something as basic as communication or sex.

We're really terrible singers, we explained.

"It's no excuse," he insisted. "You still must sing. Any song, just sing! *A-B-C-D...*"

A beat passed. Jon and I acquiesced: *"E-F-G?"*

We continued from *H*, the three of us singing into the night, the car careening down Ninth Street. We made it to *Z*, miraculously alive, and invigorated.

"Once a week, you must sing together," the driver said. "Be playful and you will stay united."

We reached the restaurant, left the car, and asked each other, did that really happen? If I were a believer I would have thought this was divine intervention. But since I'm not, I think it was a very kind, very wise, slightly nosy stranger who we were lucky to meet by chance.

My parents raised me to see love as holy, and Jon and I have always thought of our love as a kind of religion. Not supernatural or preordained but something to trust in, something to honor, something to cherish—and not take for granted.

Like any religion, our love has its hallowed origin story (the steamy August night our friendship finally turned romantic) and annual holidays (the anniversaries of that first night, of the day we decided to be exclusive, of our wedding) and those occasional, rapturous moments of transcendence. But we'd been missing another crucial element: a weekly sacrament, a regular reaffirmation of the devotion and joy at the core of what we'd built together. The thing you are obliged to do regularly, at an appointed time, to remind you of your values even when you are grouchy, busy, or annoyed. Even when you really don't feel like it. And this cab driver, whose name I desperately wish I had gotten, gave us that.

A lot has changed since that cab ride. Jon found a job he loves. We moved from New York City to Boston for it. We have a child. Yet the alphabet song is immutable. We've sung it every weekend, almost always on Saturday morning, usually upon our first eye contact of the day. We take a deep breath and hold it a moment, a kind of signal, and then begin. We sing the alphabet when we're crazy in love, when we're mad at each other, when we're rushing to be somewhere. When we're apart, we sing it over the phone.

From Southern Baptists to the shaman of Siberia, singing gets incorporated into belief. It even goes beyond our species. A vast array of birds, insects and whales perform mating rituals that revolve around song. Weekly group singing is at the heart of many religious services. And even secular rituals. How many sporting events the world over start with national anthems? How many kindergarten classes sing the same songs each morning?

The alphabet song, particularly, would not be the right thing for everyone. It's incredibly silly. Almost too silly. But over the five years since that cab ride, it's occurred to me that that's kind of the point. It's the silliness that brings down one's defenses, that creates the bond, that makes it special, that makes you feel vulnerable. That is what's essential to being part of anything bigger than yourself, like a marriage, or our whole wide universe.

Lately, I have another weekly ritual, in addition to the alphabet song, which also involves singing. On Tuesday afternoons, I take my daughter to a small rehearsal room on the third floor of a theater in our neighborhood for baby music class. She loves it. Unlike her mother, she seems to have some musical talent, or at least a sense of rhythm. Occasionally, when his work schedule allows, Jon comes, too. We sing "Twinkle, Twinkle, Little Star" and "Old MacDonald Had a Farm" and songs I'd never heard before, about shaking maracas and cleaning up. The children play toy instruments and bubbles are blown at the end. There are three other moms of girls all born within about three weeks of ours. Those other moms and I have become very close. We've formed a little community and have leaned on one another for support, for advice, for wisdom. And each week, we marvel at how our babies are slowly transforming into kids. It's hard to notice when we see our own children every day, but with the four of them together we get a sense of time passing. We also become a little more enlightened as we take time to bear witness to what is most sacred to us. For me, this is another kind of church to honor another kind of miracle. A new, small sect, we worship tiny, mysterious deities with powers and wishes we do not fully understand, but whom we love unconditionally, and hope to please with our sacrifices.

chapter three

Spring

To what purpose, April, do you return again?
Beauty is not enough.
—EDNA ST. VINCENT MILLAY, *"Spring"*

Religions have a habit of squatting on things which did not
originally belong to them.
—ALAIN DE BOTTON, Religion for Atheists

This planet we're on is a little off. The way it rotates, I mean. We are at an angle from the sun. The angle is about 23.4 degrees and it's called an axial tilt. If we didn't have this tilt we'd still have weather, but no seasons, and the lengths of days would not change over the course of the year. The summer and winter solstices are the extremes, the longest and shortest days respectively. The spring equinox is a kind of tipping point, marking the moment we are closer to the longest day of the year than we are to the shortest. As our planet makes its yearlong, oblong trek around our star, the northern and southern hemispheres get different amounts of sunlight because of our tilt, regardless of how far we are from the sun. The day that the tilt provides the most sunlight is the summer solstice in one hemisphere. That same day is the winter solstice in the other. As we make our way around the sun, summer slowly turns to autumn and winter to spring. When it's spring in one hemisphere, it's autumn in the other. North of the equator, where the vast majority of humans live, the spring equinox falls on about March 20 each year. (There is some small variation in the arrival of the equinoxes and the solstices, a day or occasionally two in either direction, in part because the calendar we use includes leap years and time zones that inch us back and forth year to year.) It's tangible and provable no matter what you do or do not believe.

Alongside the shift toward warmer weather, the return to more light that the spring equinox heralds is innately uplifting. As are the signs of new life after a hard winter. Maybe we have evolved to love spring because it signals we are out of danger, less likely to freeze or starve. Maybe seeing everything around us being reborn assuages our deepest fears about mortality. Either way, the joy of spring requires no dogma and no faith to experience.

Think of how different life on Earth might be if not for this tilt, a by-product of some still mysterious astronomical event, or maybe events. A collision in Earth's early years likely knocked us asunder. The pieces of debris that were flung into orbit during that great impact may have become our moon. Back then we may have had a much greater axial tilt, but our new moon's pull brought us the one we have now.

What an epic blessing in disguise for Earth, or rather for Earthlings. This impact has resulted in so much beauty, like the changing lengths of day, and maybe phases of the moon. It has deeply affected the everyday lives of a wide array of creatures who would not come to be for a few billion more years.

And some of those creatures (us) got very interested in how the seasons change. Keeping track of them, marking them, celebrating them. Partly this was just sensible. Knowing that the monsoons or the blizzards were coming helped us prepare and survive. Different foods grew at different times of year and that was extremely useful information. But there was an emotional element, too. We felt different feelings. If you live even several thousand miles from either pole, short, cold days elicited certain states of mind. Long, sunny, hot ones brought out others. If you lived nearer to the equator, the day lengths don't change as much, but you still have seasons and might feel very differently during the dry spells than during months of regular afternoon torrents.

The equinox is the astronomical way of marking spring's arrival. But depending on where you live, the weather patterns, or the flora and fauna, it might not really look or feel like spring yet. So maybe your spring begins when the snow melts, when you can go out without tights or a coat. That would be the meteorological way of defining it. Maybe you feel it's really spring when the flowers bloom and the baby bunnies are born. Those are biological events. Whatever it is that makes you feel like spring has arrived, there is a field of science that studies it and, by doing so, honors it. No matter the moment that crystallizes this change for each of us, the idea that the dark, cold times eventually give way to bright warmth, beauty, and plenty is at the core of spring. All seems lost, but then, somehow, we receive another chance at life.

This cycle is reflected in so many of the holidays we've created for ourselves at this time of year. Passover, which happened to be just about the only Jewish holiday my family consistently celebrated besides Hanukkah, was part of this pantheon. Not every year, but often in late spring, usually in April, my grandpa Harry and my grandma Pearl would come up to Ithaca from New York City, and my mother would make a traditional Passover meal, the Seder. We'd get out the Haggadah, which is the Passover text that lands somewhere between a hymnal, a history book, and an instruction manual. There are thousands of versions. Families choose theirs based on their level of orthodoxy and time commitment, among other personal preferences. But even with the identical Haggadah, no two Seders are exactly alike.

At our house we'd get out the special Seder plate, the good dishes and silverware. We dressed up, which I loved, being an enthusiastic over-dresser all my life. It felt special, exciting, and uplifting somehow, despite the narrative we were recounting.

So much of ritual is the retelling of stories. A philosophy requires more than just a list of things to believe in. These tenets must be illustrated in a way that moves you, draws you in. My mother's vivid, careful telling and retelling of the family sagas, origins, heartbreaks, and triumphs were part of an ancient tradition, older than the written word. I hear her voice in my head when I think of my great-grandparents sitting on the street corner or Grandpa Harry confessing his skepticism. The pictures she paints with words breathe life into real events that could be lost to the ages without her. Events neither she nor I witnessed but that live on in us. This same skill set is how religions were and are able to spread. Parents telling their children what led to them, what their ancestors were like, and what was important to their people. The key was telling them not just once but over and over in a compelling enough way that the stories would be remembered and retold.

A traditional Seder tells the story of how the Jews escaped from slavery in Egypt by way of the Red Sea. Needless to say, it's usually focused around Yahweh, the Hebrew God, who plays a central role in the getaway. My mother's favorite part is the singing of the song "Dayenu." It's really catchy, the kind of song that gets stuck in your head. But that's not why she likes it. It enumerates miracles, blessings, and gifts given by God to the Jews in the Old Testament—parting the sea, drowning oppressors. After each act of God, "Dayenu" is sung, which roughly translates to "That would have been enough." As in: "If he had just taken us through the sea on dry land that would have been enough" . . . "If he had just given us the Torah that would have been enough." My mother doesn't believe Yahweh guided our ancestors out of Egypt. She likes it because it's a song of appreciation. We've often talked about what it would be like to have a secular version of "Dayenu." *If he had only given us the sunshine, that would have been enough. If he had only given us the flower blossoms, that would have been enough.*

That only leaves the question of who the "he" is here. Maybe the "God of Spinoza," the one Einstein once described as "reveal[ing] himself in the orderly harmony of what exists." In other words, the God of the physical laws of the universe, a kind of literary—but not literal—personification, along the lines of Mother Nature. I have always loved this idea.

Besides singing, Passover features special reading aloud,

symbolic foods like an egg, and the ritual dripping of wine dots on our dinner plates. Also, opening the front door to invite in a theoretical and symbolic stranger named Elijah. For me, the highlight was the hunt for the afikomen, a piece of matzo usually wrapped in a napkin and hidden around the house for children. When you find it you get a little prize, maybe a five-dollar bill.

The whole story of Passover is dark. It's classic Old Testament God. We're talking slavery, the systematic murder of small children, doorways covered in lamb's blood, plagues that include—but are not limited to—lice, boils, locusts, and really bad weather. Also a whole bunch of guys drown, but they're the bad guys, so that's actually one of the good parts. And it ends with the Israelites being lost for four decades in a dry, hot hellscape. It's the stuff of horror movies. And yet, Passover has a positive, festive vibe.

Around the same time of year, Easter weekend begins with Good Friday, the commemoration of a gruesome, terrifying public execution, Jesus's crucifixion. It's heart-wrenching to even imagine. But Easter Sunday marks Jesus's return, a miracle and a cause for celebration similar to Passover.

The spring equinox is almost the same day each year, but Easter and Passover are not as reliable. They move around. Some holidays stay on a fixed date in the Gregorian calendar (the one the vast majority of people on Earth use), like

December 25, for example. But Easter, like all Jewish holidays that still follow the Hebrew calendar, changes year to year. In the very beginning, the first Christians celebrated Easter on Passover. Passover predates Easter, because Christianity is an offshoot of Judaism. The first Easter celebrations coincided with Passover because, according to the New Testament, the Crucifixion and Resurrection took place around Passover. (There is some scholarly debate about whether or not Jesus's Last Supper might have been a Seder.) As Christianity spread, people marked the Resurrection on different days in different regions. In 325 CE, when the Roman Empire was formally transitioning from the polytheism of the past to the monotheism of the present, Emperor Constantine I got some guys together to make some big decisions. There, in Byzantine Turkey, at the Council of Nicaea, they established how the church would manage its money, its hierarchies, and the sex lives of priests. They also decided to unlink Easter from Passover, possibly because the Jewish calendar was not steady or uniform enough. It was also a way to let the new holiday stand on its own. Easter was set for the first Sunday after the first full moon after the vernal equinox. (Despite this push for unity, Eastern Orthodox Christians still mark Easter on a different schedule, tied to the Julian calendar of ancient Rome.)

Perhaps because my mother understood that these events were all wrapped up together, she and I decorated

Easter eggs with food coloring every year around the time we planned our Seder. Not until much later did it occur to me that the eggs were a Christian tradition. It didn't seem Christian to me. It didn't have anything to do with Jesus or the Crucifixion or the Resurrection or even God. It just seemed like art that reflected all the new life popping up outside. I didn't know this at the time, but eggs have always been prominently featured in creation myths, from ancient Egypt to the Hindu Vedas, maybe even back to the art of the Paleolithic world. The idea that life—maybe even all of life—emerges from an egg seems to have made sense to people for eons.

I have a friend on social media who, every spring, posts pictures of the gorgeous eggs she's painted to look like the galaxies of deep space. She drinks nature-themed cocktails while she decorates, and watches an episode of my parents' show, *Cosmos*, while the eggs dry. She describes herself as secular but spiritual, and has successfully combined a religious tradition with an understanding of the universe as revealed by exploration. It moves me in a strange way, like seeing an artifact from a society where science and religion were never at odds.

As with so much of culture, we adopt the part we like, the part that speaks to us. Sometimes this is theft or appropriation, but sometimes it's an homage. A reference to another time, now incorporated into something new, something still relevant. Without this kind of co-opting, so many ancient ideas would be lost forever. I find it stirring and lovely that themes and symbolisms have managed to stay with our species for millennia.

The first time I ever took Jon to a Seder, we were living in London. We went to the home of my family friends, one of whom had lived through the blitz as a little girl and had taught her brother to read while they hid from the Nazis' bombs. Jon was very game, as he always is when I ask him to try something new with me, but had no idea what to expect. It turned out to be a very unorthodox Seder. Children's songs were repurposed to tell the story of Exodus. The tune was "She'll Be Coming 'Round the Mountain," but the lyrics were "He'll Be Comin' Down the Mountain." "Take Me Out to the Ball Game" became "Take Me Out of Egypt." It was more fun than Jon had been expecting. He happened to mention to some of the other guests that it was his first time celebrating Passover. They looked at him in his borrowed yarmulke quizzically. "Your name is Jonathan, you come from New York, you work in finance, and you're not Jewish?"

"No," he said.

"Are you sure?" they asked.

"No, I'm not sure. How could I be sure?"

I loved this. And I felt delighted at the idea that Jon might secretly be Jewish for some sort of sick, incestuous reason I cannot, to this day, rationally explain. But I had a recurring fantasy about some long-ago relative of Jon's settling in France during the Diaspora, keeping his identity secret and learning to pass. This kind of story was the very reason I still celebrated Passover, and clung to other select bits and pieces of my Jewish identity. At its core, Jon's joke was about the fact that people who were Jewish had to sometimes pretend not to be Jewish. Because they were forced, because it was dangerous, because it was the only way to survive. They thought they'd bury away their true identity forever, and by losing some part of themselves could save the lives of their children.

Years after that first Seder, I asked Jon how he felt and what he thought that spring night in London. He said it had reminded him of a cross between Thanksgiving and Easter. He loved how thought-provoking it was but couldn't help feeling a little left out, a little awkward. He doesn't have much experience being the minority, and this was a new, strange feeling for him.

In his joke Jon said, "How could I be sure?" But that was in 2008, before DNA tests that mapped one's entire ancestry were readily available. Now that they are everywhere, I find I have a constant stream of tall, blond acquaintances and friends telling me they've just discovered that they are a little or a lot Jewish.

I'm not completely sure if this was in my mind when

we decided to buy a DNA kit for Jon's dad for Christmas. Lo and behold, the secret code in my father-in-law's saliva revealed that he is 7 percent Jewish. Maybe there was once some person in his lineage who had moved away, changed their customs and their name, thinking no one would ever find them out. Never dreaming an invisible molecule in the very fiber of their being would someday, posthumously, give them away.

Judaism is matrilineal, so regardless of Jon's genes, our daughter is considered Jewish, at least technically. She'll have to decide for herself, of course, how she identifies, but I will share with her some version of what her maternal ancestors have been doing each spring for six thousand years. And also the Easter eggs. Not only because her ancestors on her father's side have nearly all been Christians, but because ritual and art are so deeply intertwined for me. The cooking of food and the singing of song are art forms, and I love those. But the small visual art projects take me back to the weekly construction paper worlds with my mom. And to every majestic church I've ever stepped foot in, to every soaring minaret I've ever seen, to every fertility goddess figurine behind glass, every object of art created to convey a belief. They are all still hauntingly beautiful to this non-believer.

Anyone who's ever visited a museum knows there is so much one can draw from religious art even if it's not the art of your religion. This goes for holidays, too. There is so much I love about Passover that requires no belief. In recent years, Jon and I have started using something called *The Liberated Haggadah: A Passover Celebration for Cultural, Secular and Humanistic Jews* for our Seders. It was created by the secular rabbi who married us. It's not so much a book as two copies of a stapled packet I keep meaning to get bound or laminated but haven't. It does not mention God but instead celebrates life, wine, and freedom. The emphasis is on the history of the Jewish people, the way we've managed to continue living despite a long series of setbacks, and the fact that other people—people of a wide variety of faiths, races, ethnic backgrounds, and sexualities—have suffered the same heartbreak, violence, prejudice, and destruction. The nontheistic version of the holiday goes something like this: Our people were slaves and that was horrible. We're lucky to be free. But right now there are others who are enslaved. Physically and metaphorically. By poverty. By discrimination. By violence. We must help them. This modern Haggadah lists different plagues, among them: hunger, illiteracy, pollution, poverty, racism, violence, and war.

The last few years Jon and my Seders have been less elegant than the ones my mother put together when I was small, with her wedding china and my father and grandfather in suits. The other couples who attend our Seders are almost all also "mixed marriages," both partners generally secular, one raised vaguely Christian and one raised vaguely Jewish. But both eager to mark changing of the seasons.

Jews and Christians are, of course, not the only people celebrating spring.

Before Islam arrived, the people of the Persian Empire were majority Zoroastrian for millennia. They believed in a single God, the Wise Lord, who is powerful but not omnipotent. Nowruz was their New Year, literally "new day." Today Iran is a strict Islamic theocracy, but Zoroastrianism still echoes in the celebrations of its citizens. Nowruz is so beloved that even the ayatollah tacitly allows this pre-Islamic holiday, which also features the dyeing of eggs.

In Egypt, Sham El-Nessim too calls back to a pre-Islamic world but is nonetheless included in the modern calendar. It had been celebrated on the spring equinox from at least as far back as 2700 BCE, but during Egypt's Christian period it was moved to Easter Monday, perhaps to keep the holidays from competing, where it remains. In the modern era the sacrificing of foods to the gods has been supplanted by picnicking, but the decorating of eggs has remained.

Chinese New Year is one of the most widely celebrated holidays on Earth. It falls well before spring, between the end of January and the beginning of February. Winter, really. But despite what the calendar or the thermometer says, it's often called the Spring Festival. Giant dragon floats, countless red lanterns, and fireworks are employed in the celebration of good luck and the hope that the coming year will be better than the last.

Other spring festivals seem profoundly modern. But even the most envelope-pushing, scandalous stuff is about human biology at its core. In modern Japan, on the first Sunday in April, as the famous cherry trees blossom, Kanamara Matsuri is celebrated in Kawasaki, just outside of Tokyo. It's the Festival of the Steel Phallus. We're talking about tens of thousands of people gathering for a parade of giant penis-shaped shrines, plastic penis noses, and brightly colored penis lollipops. And fund-raising for HIV research. It's the picture of a modern, progressive, sex-positive holiday. But it's linked to a shrine where, between the seventeenth and nineteenth centuries, sex workers would pray for relief from sexually transmitted diseases. Kanamara Matsuri appears to be rooted in legends of possessive demons that once hid inside vaginas and jealously bit off penises during sex. They were eventually defeated with the help of a blacksmith who fashioned a scabbard—the steel phallus being celebrated each April.

Liberalia was a Roman spring fertility festival, and like Kanamara Matsuri, it involved giant fake penises that were paraded through the streets. It fell on March 17, marking the day when young men would shed their boyhood clothes and first don the "virility togas" of manhood. Some elements of Liberalia, like so much else in Roman culture, were inspired by the Greeks, who held an array of festivals in honor of Dionysus, the god of wine. One, Anthesteria, or the "festival of flowers," kicked off spring with wine-tasting and sacred, secret rituals performed by women.

As a kid I had a copy of *D'Aulaires' Book of Greek Myths*, a large yellow paperback featuring an illustration of an expressionless god, with rays of light streaming out from his head in all directions, driving a chariot pulled by four white horses. This particular god was called Helios, and he was a sun god. According to the people who flourished on the Greek Isles between about 800 and 146 BCE, Helios rode his chariot across the sky each day, bringing morning to the Earthlings, driving out the darkness. (So he was really the personification of the rotation of the Earth, since it is us who moves around the sun, not the other way around.)

I adored this book, and I frequently demanded my parents read it to me when I was not yet able. Later, I would read and reread the same stories myself, memorizing the genealogy, peccadilloes, and particular powers of each god. They were depicted here as blonds, mostly with blue eyes and features that, in retrospect, seemed more representative of Odin's family than Zeus's. I pictured ancient Greek families worrying and wondering, living their lives. I tried to imagine them, meditating on the idea that they were once *real* people who *really* lived. And now the world they inhabited had completely changed. From this obsession I learned a little Greek history and geography. But mostly I learned that religions come and go. I marveled at the intricate beauty and sorrow of these tales and the idea that these mighty gods had lost their legions of devotees. I sort of felt sad for them, but at least they got to live on in the names of sneaker brands and space missions.

It's perfectly acceptable to see those gods as reflections of their believers. No modern middle-school history teacher suggests Zeus was actually born secretly in Crete after his father, the titan of harvest, devoured all his older siblings. That doesn't get presented as fact the way, say, the whens and wheres of the Peloponnesian War do. We learn that the story of Zeus is just what the Greek people *believed*, and it's implied that they were mistaken about the true circumstances of the universe.

We learn that Demeter was their goddess of grain, because grain was important to the Greeks. Grain wasn't important to the Greeks because Demeter was looking over them. The ancient Egyptians and the Mayans worshipped sun gods because our nearest star played such a vital, double-edged role in those sweltering climates. Same with Freyr, the weather god of the Norse; Mars, the war god of the Romans; and the fertility goddesses from Tonacacihuatl of the Aztec to Mbaba Mwana Waresa of the Zulu. The rains, the crops, the invaders, the kids—this is what determined our future in an uncertain world. This is what we worried about, so these are the gods we created.

There is no taboo against saying these gods were invented to cope with the dreams and anxieties of their ancient disciples.

So I read my book of Greek mythology as literature instead of history. Which was not so dissimilar to the way we looked at the Haggadah. This is what our ancestors believed. There is wisdom, insight, and poetry in it, but it's not what we believe.

One story from the *D'Aulaires' Book of Greek Myths* stayed with me powerfully: the story of Persephone. She was the daughter of Zeus and Demeter, who actually oversees more than just grain—she guides the whole of agriculture and reproduction, among other things. But beloved Persephone is abducted by Hades, god of the dead, the underworld, the devil of ancient Greece. Demeter is beside herself. She leaves Mount Olympus for Earth disguised as a lowly old human, searching tirelessly for her daughter and letting her role as the grower of grain fall by the wayside. Food becomes scarce. People starve. The other gods become concerned. Eventually they decide it's worth their while to band together to organize Persephone's rescue. When Persephone arrives back safely in the arms of her mother, the world blossoms, food is plentiful, and the humans are saved.

But while in the underworld, Persephone consumed some pomegranate seeds, eternally connecting her to the realm of her kidnapper and obliging her to return to the underworld every year. And the rest of the world is, in turn, trapped in a cycle of barren winter following plentiful spring forever. For the Greeks who believed, this was the origin of the seasons.

All these spring legends are about suffering and heartbreak giving way to joy. Each contains a secret, a hidden miracle, offering hope when all seems lost. This is spring itself. The themes of renewal, rebirth, resurrection, and rescue from death are not religious ideals in conflict with nature but rather rituals inspired by the biology of plants and animals.

When I was growing up, Passover wasn't the only springtime holiday we kept sacred at our house. My mother invented a "first fruits" festival just for me: Blossom Day. As the long upstate winter finally faded, I would start to make daily examinations of the dogwood tree outside our dining room window. My eyes would play tricks on me. *Is that something? A teeny tiny hint of a bud? No ... Maybe tomorrow.* I would wait, day after day, until there was, indisputably, a blossom on the dogwood tree. My mother and I would go out and marvel at it. And then plan something like a tea party. Usually it was just the two of us, maybe a girlfriend of hers who lived nearby. Gifts were exchanged, but the biggest gift was the idea that celebrations could be invented, that we could choose to honor what was most meaningful to us, that and the knowledge that spring was finally here. Blossom Day stripped away all the layers to create a new holiday. Some spring day, we will stand at that same dining room window with my daughter and bask in the wonder of it: the tree that seems dead but isn't. Something new is growing. The weather will be better soon. The days will be longer. The sunshine is coming. It still feels magical. Miraculous, even.

*

chapter four

A Daily Ritual

To be ignorant of what occurred before you were born is to remain always a child. —CICERO

The heavenly bodies, the earth, the river Nile, its water, mud and slime, man, animals, plants, reptiles, insects, all the forces of nature, animate and inanimate, were sacred, and were partakers of divine honors . . .
—ANN ELIZA SMITH (AS MRS. J. GREGORY SMITH),
From Dawn to Sunrise: A Review Historical and
Philosophical, of the Religious Ideas of Mankind

Jon and I have several small daily rituals. Every morning, he wakes up first, makes coffee, and brings me a cup in bed. Then I thank him and tell him how wonderful I think he is. This small act of romance sets the tone for our day and, in turn, our life together. Later, when he leaves his office, he texts me, "On the way!!!" This still gives me a little thrill. Like the feeling of knowing you're going to see your crush. And this too is a little ritual. In a sense, everything we do a particular way that holds meaning for us is a ritual. Especially when there are other ways it could be done. He could just come home without sending a text, but by letting me know he's coming I get to enjoy that giddy sense of anticipation.

We all have these rituals in one way or another. Maybe it's the route you take to work or the way you prepare dinner for your kids. There is a little narrative that goes along with so many of our daily tasks. When I moisturize my face before bed, I imagine the legends of the Fountain of Youth. That's the story being told in every ad for anti-wrinkle cream. These small rituals give us comfort and offer a kind of rhythm, a reliable pattern, and, I think, an artificial sense of certainty. If I can really slow the hands of time with a skin-care routine it will be because of science, not magic—if we must delineate between the two, although doing so can rob us of the thrill of both. When my daughter and I leave the playground or some other place frequented by small runny noses, I ask her if she's ready for the magic potion we put on our hands to protect us from sicknesses. Antibacterial gel is not usually the stuff of fables, but it could be. Imagine encountering a sect somewhere who devoutly carry small bottles of clear fluid around with them and believe wholeheartedly that rubbing the contents on their hands shields them from danger. We would think they believed in magic. Why don't we? Just because we know why and how it works? Why does the provability of something rob us of the thrill of it? Even the coffee Jon brings me every morning feels like sorcery. Something grows in the earth. It's harvested, roasted, ground, and percolated. I drink it and, like Alice, I am changed. It wakes me up. It gives me strength and speed; superpowers, really. At the end of the day, a glass of wine is another kind of potion. Another plant comes out of the earth, is readied by technique and time, and when drunk has the power to unwind you, let you take a step back, and quell stress. So many everyday rituals amount to a magic trick being performed by biology, technology, or some other branch of science. How different would our daily lives be if we found ways to celebrate even the smallest wizardry in life?

No matter how many rituals we have, we all also experience uncertainty every day. Maybe even every hour. *What should I do? Is this true? What is going to happen?* Impossible to predict, the future tortures us. Something absolutely shocking can happen at any moment. What a toll this takes on us. How many small tools have we created to combat this, to assert some control in the universe?

Jon's dad told me that when Jon and his sisters were small, he had a superstition: for every kiss he gave his children, they would live another day. He would cover them in enough smooches to make them Methuselahs, not because he truly believed that was how it worked, but because it gave him a sense of influence over his deepest fear: that something bad might happen to them.

Our deepest fears power so much of what we do. I was a cerebral, sensitive child. I spent a lot of time wondering about the dioramas at the museum and the whereabouts of my late paternal grandparents. The realization that everyone I knew, myself included, would someday die sent me into a kind of existential crisis. I recovered, but I was obsessed by the idea of death, and I remain so. Especially the fact that it's inevitable and often unpredictable. This ambiguity haunted me, so while still small I devised a plan, a kind of personal superstition. Every evening while being tucked in I would say to my parents, "Don't forget! Don't die!" and every evening my mother would say, "I promise!" and my dad, accuracy zealot that he was, would say, "I'll do my best!" I don't know if it was because he was already in his fifties by then and had faced several serious health problems, or just that he wouldn't want to break a promise by falling victim to something he could not control. But this was our nightly ritual, and for a time, it gave me the desired illusion of control, safety, and certainty.

It's easy to forget that even the ending of a day is an astronomical event. Before we understood what made a day, it seemed like it was a function of the sun appearing, moving across the sky, and then disappearing into the darkness. We talk about sunrises and sunsets, despite it being the Earth, not the sun, that is moving. The misnomer aside, this pattern has a profound effect on us. A day is twenty-four hours because that's the time it takes for the Earth to rotate on its axis once. Our biology, our circadian rhythms, link us to the way we move in space. For most of our history we had no clocks and no calendars, but we always had night and day. There is something profound about doing something, anything, to mark this rotation. And far and wide, we have.

On important days, there is a ceremonial chant among native Hawaiians called E Ala E. It is directed eastward before first light, asking the sun to rise. In India, Nepal, and beyond, a small, ancient ritual of throwing cow's milk on a fire at dawn and dusk has been performed by the priesthood class for thousands of years.

The goal of so many daily religious rituals is to tap into our sense of gratitude for the great and powerful force to which we owe our lives. This is no less important if you think those forces are physics and biology. Yoga and meditation are rooted in religion but have taken on secular lives of their own.

A quiet moment at the start or close of a day is a ritual with an infinite number of private permutations. For many religious people, prayer is a crucial, defining staple used to gain clarity, get in touch with what you believe, what you wish for and aspire toward, another way to face the uncertainties of life. I spent much of my early life with one such religious person.

When I was six weeks old, Maruja Farge came to live with us. She was my nanny and became a member of our family. She had a profound influence on my life. She was also a devout Catholic, a true believer.

When Maruja was a teenager in rural Peru, her mother died giving birth to a younger brother, who also soon died. This was the late 1930s, in a world where single men rarely raised children. Just as Rachel's father had, Maruja's father sent her sisters to live with an aunt. But Maruja was the eldest, so she was sent to a convent in the Andes Mountains. There she became a cloistered nun. For almost twenty years her only contact with the outside world was a kind of turnstile through which villagers could leave anonymous gifts, maybe a chicken or a bag of quinoa, as a kind of tithing for the nuns. One day it was Maruja's turn to receive the donations. She spun the divider to discover a newborn baby. As my mom remembers, having heard the story from Maruja decades later, Maruja's heart raced as she held the newborn in her arms. The Mother Superior took the child away. His or her fate is unclear. But the experience put a seed in Maruja's mind. Suddenly, she felt a pull toward the outside world. But, devoted to the Church, she did not want to break the vows she'd sworn when she became a nun. Over the next few years she wrote letters to the Vatican requesting official permission to start a new life. Finally, she was given the papal blessing to come down from the mountains and out into the world.

Maruja was in her thirties by then and, according to her, an "old maid," beyond the age at which she would be a desirable wife. So she became a nanny (in my personal opinion, the greatest one of all time) and raised more than a dozen children all over the world, including me in our secular, scientific, Jewish household in upstate New York.

For the eight years she lived with us and for the many subsequent summers she would spend visiting us, she was completely open and straightforward about her beliefs. I knew exactly what she believed and I knew that it was different from what my parents believed. My parents weren't afraid that exposure to other belief systems would somehow harm me. The more I knew about what people thought, the better off I was. Once, early in my fascination with death, I came to my parents with a question:

"Maruja says when you die you go to heaven and there are angels playing harps and you're with God. And you guys say it's like you're asleep forever with no dreams. Who is right?"

My parents, without missing a beat, said in unison, "Nobody knows!"

And they didn't just say it. They announced it like good news, joyful, enthusiastic, beaming.

This exchange was revelatory for me. Not because it gave

me any clarity on the mystery of death, but because it gave me a window into the nature of life. It taught me that there is no shame in not knowing. Uncertainty is real. It need not be glossed over or buried. We can embrace it, even while we try to understand what we can.

The pre-Socratic sophist philosopher Protagoras of Abdera wrote this: "Concerning the gods, I have no way of knowing either that they exist or that they do not exist, or what they are like in form. There are many things that impede that knowledge, both the obscurity (*of the subject*) and the brevity of human life." That's kind of my position, too. (Right after he wrote that, the Athenians outlawed teaching that kind of heresy. Presumably they felt threatened somehow by Protagoras's ideas.)

In the 1950s, the philosopher Bertrand Russell made the argument that the burden of proof must be on the believer, not the skeptic. He did this by asserting that there is a teeny tiny teapot in space that cannot be perceived by even the most powerful telescope. He suggested that if this idea was supported by ancient books, it would be deemed more likely, even if the ancient books included no proof. We cannot be absolutely sure that there is no space teapot; it's possible there is. But we don't have any reason to believe in it.

This approach applies way beyond theology. My dad was frequently asked by interviewers and fans if he "believed" in the existence of extraterrestrials. "I don't know," he would say "I don't have enough evidence." This would sometimes frustrate people, and there were a lot of follow-up questions about his gut or his instincts. My dad really wanted to know if there were aliens, and if there were, he wanted to know everything about them, but he refused to let his deepest wishes fog his judgment. He was a devout scientist.

In his book *Pale Blue Dot,* my dad wrote, "Science demands a tolerance for ambiguity. Where we are ignorant, we withhold belief. Whatever annoyance the uncertainty engenders serves a higher purpose. It drives us to accumulate better data."

When a new person discovers my good fortune of being my father's daughter, they often ask, "Well, are you a scientist?"

I don't know how many women go into their fathers' businesses nowadays, but I did not. And yet some strange part of me wants to answer *yes.* Not because I hold a degree in anything more objective than dramatic literature, but because my parents instilled science in me not as a job but as a worldview, a philosophy, a lens through which to see everything. Just as not every Catholic is a priest, not every adherent to the scientific method is a scientist.

The discussions that fueled my parents' workdays flowed over into dinnertime, and that enriched me enormously. They managed to explain even the most complicated concepts clearly and without ever talking down to me, with a kind of gentle intellectual respect, as though I was some kind of tiny professor trapped in a little girl's body. I think it's this same skill that made their work accessible to so many non-scientists around the world.

During those dinnertime discussions the greatest thing I could do, in the eyes of my parents, was to ask a question to which they did not know the answer. To them it was the sign of a curious intellect scanning the horizon for new mysteries, which might lead to new understanding. I never once got a "That's just how it is" or a "Because I said so." Instead, my father would, smiling, retrieve a volume of our brown and green *Encyclopædia Britannica* from above our sort of kitschy, 3-D, glass-and-neon painted model of the Milky Way galaxy. Together we would look up the answer and revel in a shared clarity.

Sometimes a different reference book was called for. I can still see him standing on our ottoman to get a huge atlas down from a high shelf to resolve some geographical query. Sam, then less than two, gazing up at him and, with a kind of respectful reverence, declaring it a "big book!"

For many, reading passages from their holy text offers daily enlightenment and answers to the endless questions that run through our minds. For us, it was the encyclopedia, atlases, and dictionaries. And like finding a new verse in a sacred book, or reading an old one with fresh eyes, this enriched me emotionally, not just intellectually. Every day I gained a deeper sense of the workings of the world and the universe. Every day I got a little closer to a sense of understanding. Learning to draw connections thrilled me. It made the world less overwhelming. I felt more confident, braver. The more I learned about history, the more I understood about geography. The more I knew about etymology, the more Latin I could piece together. In those days, every encyclopedia entry I read felt like it was snapping in a new puzzle piece. I thought I might soon see the entire picture.

But as I grew, I realized the puzzle had no edges, no borders. It went on forever in all directions. Every new piece just revealed how many more pieces were still missing. I came to understand that I could never get to the complete picture.

So the metaphor changed. Instead of a puzzle, being curious became more like being a collector of small, beautiful objects, of which there are a seemingly infinite number on Earth, like seashells or stamps. I'd never have them all, but each new kernel of understanding was like a new, gleaming gem. When I learned something, I imagined putting it in a special box with my other treasures, seeing how they went together, how they matched or clashed. Learning became addictive, an obsession. Soon the urge to try to answer any lingering question was overpowering, each answer in turn eliciting another question, some parochial, some cosmic.

The parochial ones are often maligned as "trivia," but every piece of trivia is a small clue to something else, a glimpse at how we fit into the universe.

Everything became fodder for this ritual. I'd find myself somewhere new and start to wonder, What is the name of where I am at this moment? The street? The neighborhood? The city? The nation? These proper nouns could have been anything, but this place has its name for a specific reason. Was it named after a person? Another place, picked by some homesick explorer? Is it a ridiculous Anglicization of an indigenous word? A slow contraction of words over centuries? If New York is named for York, England, where is Zealand? Where did the word *America* come from?

Soon it became clear to me that there is a second layer to everything. Often there's a third and a fourth and a fifth, subtext upon subtext, subtle reference within subtle reference. The best literature, movies, and art had the most layers, I thought. Finding these felt like being a detective.

I think I was somewhere around five years old when I asked my parents about Maruja's view of death. I was in that endless "Why? Why? Why?" phase, something was changing in my brain. I was thirsty for knowledge. My neural connections were maturing. My synapses had been forming at record levels, making more than I'd ever need. The extras were then "pruned" away as I grew. In developmental psychology this period is sometimes called the "intuitive thought substage." In Catholic canon law it's called "the age of reason." In both cases, it's when tiny humans between the ages of four and seven start to use basic logic. Our species doesn't have as many rituals to celebrate this as we do for birth or puberty, which bookend childhood. There are some, though. For young Catholics it's time for First Communion. In Japan, girls aged three, five, and seven, and boys aged three and five, get dressed up in traditional clothes and go for a visit to the shrine on Shichi-go-san, an annual holiday that celebrates the growth and development of young kids. If you happened to be an ancient Spartan boy, age seven was time to enlist. And today, for most cultures on Earth, usually somewhere between ages three and six, children partake in another rite of passage. They dress up in new clothes, are given a special bag to wear on their back, take pictures, and receive kisses, before being picked up in a large, usually yellow mechanized conveyance and taken to the local institution of learning. The first day of school is, at heart, a celebration of a biological change. Children are big enough to be off in the world, their brains and bodies ready to absorb new information and start down their own path.

For about another decade or two, depending on the person, this daily ritualization of learning called school continues. We are, in some way or another, offered something new to understand, memorize, or question. Sometimes it's aggressively boring, sometimes it's inaccurate, occasionally it's inspiring, but it's always a chance at new enlightenment. If we are lucky enough to have a few great teachers, the exploration of new ideas becomes a gratifying routine. Eventually we either graduate or drop out, and have to decide how we will keep it going, out in the world, without the structure of syllabi and testing.

In school everything is divided. Civics, the history of the world, literature, art, languages, math, and science are sectioned into different categories. The bell rings out to say, "That's over . . . ! Now here's this totally unrelated thing taught by a different person in a different room!" But that felt artificial to me. It was really all one subject, one story: how human beings understand themselves and the universe. And the more connections I could see, the more interested I became in all of it.

You sometimes hear people say, "Children are born scientists." This is true in that when we are little we are full of questions and wonder, but there are some significant differences. If little children were really born scientists, they would be better at employing the scientific method. Imagine a four-year-old observing some strange thing and then formulating a hypothesis, devising a controlled experiment, observing the results, looking at the data, and revising their previous position. It would be adorable and impressive, but surprising.

Another way in which small children are not very scientific is that they tend not to consider the source. They usually take anyone in a position of authority at face value. It's very hard for them to imagine that grown-ups might be misinformed, biased, or have some ulterior motive. This is part of what makes children so vulnerable and so easy to confuse or indoctrinate. It's also one of the hardest things to learn. It was hard for me. After all, even the *Encyclopædia Britannica*, our sacred text, was riddled with outdated information and prejudice. It told the male, straight, white, Eurocentric history of the world, from the vantage of the end of the twentieth century. It was the *Britannica*, after all, not the *Encyclopedia Earthtannica*, let alone the hypothetical *Encyclopedia Galactica* my dad yearned for in *Cosmos*.

When our little girl is big enough to ask questions, we'll have a lot of places we can look for answers, some more reliable than others, some more convenient. For her, and for all the children of the twenty-first century, navigating the vast array of information available will bring new challenges. Discerning what sources are accurate, figuring out who has ulterior motives, or who is misinformed, will be difficult, but it's still better than not having access to any answers at all.

My dad didn't have any encyclopedia as a child. One of the defining moments of his life was the discovery that he could go beyond what his village elders knew. This was retold by him in *Cosmos*. He rarely ventured beyond his neighborhood of Bensonhurst in Brooklyn. It was his whole universe. When he started to wonder about what the stars might be, no one he knew had a satisfactory answer. His parents, Rachel and Sam, were poor and undereducated. They didn't know. But they knew where he could find out. His mother took him to the public library. I picture my dad as a little boy, barely able to see over the librarian's desk, asking for a book on stars. She came back with a book about Hollywood. Not what he was after. But he explained, she got him the right book, and suddenly he was on the path of curiosity that he would occupy for the rest of his life.

My dad died in 1996. He never owned a cell phone. He never had an email address. I often daydream about showing him a smartphone. I imagine telling him that this little rectangular machine contains all twenty-some volumes of the *Britannica*, also the collected works of Shakespeare, and a world atlas. You can use it to listen to all the music and read all the books your heart desires. It gives you live weather reports, breaking news, and the power to communicate in Albanian or Urdu. You can use it to see the opinions and vacation photos of anyone willing to share them, anywhere on Earth, with just a few taps. He would have loved it.

My dream is that, when my daughter is older, her exploration of the world—its history, its art, its creatures and their ways, its place in the universe—will not end when the traditional yellow bus brings her home each day. I hope that after school, on the weekends, on summer vacation, she will—as I did—see the looking up of things as a holy ritual performed by our family every day. And that somehow we can teach her to get comfortable with the idea that even with so many answers we still know so little. As with love, it's our vulnerability that opens us up to something deeper. Our willingness to be wrong, to let go of our predictions and preconceptions, clears the way to more than we could have otherwise imagined.

There are some mysteries to which we will never get the answer. We might not live to learn what came before the big bang. We won't know the eventual fate of our species. And there are answers that we will get. Now, both my father and Maruja have the answer to the question I posed all those years ago. And someday each and every one of us will, too. But until then, there is so much else to learn and celebrate between each sunrise and sunset.

*

chapter five

Confession &
Atonement

*A man should never be ashamed to own that he has been in the
wrong, which is but saying in other words that he is wiser today
than he was yesterday.* —ALEXANDER POPE

Sometimes on Sundays, when I was small, usually when
we were traveling, Maruja would take me to church
with her. I loved it. I would dance in the aisle even though it
was not the kind of church service where one was supposed
to dance. But no one seemed to mind. I was not totally clear
on what was being said, but I liked the feeling of being there.
These churches were generally grand, with high ceilings and
full of lots of friendly people in their Sunday best, all things
that very much appealed to me—and still do.

I noticed that these churches all had a little room where
people would enter one by one after the main event. This

was interesting to me. What were they doing in there? Maruja explained to me, in Spanish (we only spoke Spanish to each other), that they were confessing their sins to a priest, and that she too went in to confess her sins sometimes.

My main reaction was a desperate wish to avoid having to tell a strange man that I had, for example, had an unwarranted tantrum in the grocery store. And I stand by that. But confessions need not be made to a priest. Recognizing that we have made a mistake, acknowledging it, attempting to make amends, or at least trying not to do it again, is the pathway to growth, whether ritualized or not. Working to improve ourselves isn't only good in the abstract. It's a massive evolutionary advantage. If we couldn't learn that this plant is poisonous or that river has a strong undercurrent, we could die. If we couldn't learn to work out our differences with the other members of our community, that might kill us, too. We've always had to learn the error of our ways and strive to be better. It's just the definition of *better* that changes with time and place.

Ritual purification, both spiritual and physical, is common to many religions. It often comes from the idea that human bodies are dirty, innately flawed in their functionality and earthliness. I don't see it that way. I think the parts of us that bleed and orgasm and eat and sweat are sacred, too. It's all part of the astounding, intricate machinery of being alive.

I do agree with the premise that there is something wrong with us, something we must work to overcome. It's our xenophobia, authoritarianism, and violence. When we lived in small bands of only a few dozen people there might have been some selective advantage to these terrible qualities. But now we live in a tribe of seven billion. We are, more than ever, all in it together. Science and technology have allowed us to see one another's lives, to speak one another's languages, to learn one another's customs. They've also given us a view of our small world in the vastness. This should breed kindness. As my dad once said, "If a human disagrees with you, let him live. In a hundred billion galaxies, you will not find another." I believe our cruelty toward one another, not sex or love of knowledge, is our original sin. It's that for which we must really atone. In small instances as well as large ones.

Years ago, at a dinner in New York, I overheard a woman I like and admire talking about tarot cards or crystals, something popular but unscientific. She wasn't talking to me, but I interjected, "Oh, come on! You don't really believe that!" too loudly, from across the table, and proceeded to enthusiastically pick apart what she was saying. I thought it would be a fun intellectual exercise, a playful sparring match, but suddenly I realized she was not having fun. I was humiliating her. And everyone there was painfully aware that I was a massive jerk.

I must have been feeling very insecure that night, probably

about something wholly unrelated. I am of the firm belief that people are only really ever mean when they feel bad about themselves, projecting or overcompensating. And I was being mean. I should have apologized right there in the moment, but I was too cowardly. The next morning I texted someone we know in common. "I think I really offended her last night." And was met with a clear, un-sugarcoated confirmation that yes, indeed, I had. I had a strong urge to put the blame elsewhere. Maybe I could convince myself that she was being "too sensitive" or that I wasn't wrong because everything I had said was technically true. But deep down, I knew it wasn't a matter of what I had said, but how I had said it. I had shown her no respect and no common decency. I cringed at myself for hours, unsure of what to do. The feeling of discomfort, knowing I had caused someone else hurt, felt biological. We humans evolved to live in groups, to share the work of hunting and gathering, of keeping watch and looking after little ones. It's not just nice to get along, it has been central to our survival. After wrestling with what the appropriate restitution might be, all I could come up with was to write her and apologize, offering to take her out to lunch. She accepted, and a few weeks later we met in the West Village. The meal was initially awkward but ultimately lovely.

I hope I've grown from that experience. I think I have. I certainly learned that it's often worse to envision admitting you're wrong than to actually do it. I also learned that a pre-existing, formalized framework for apologizing would be nice. Lots of different religions offer this to their followers, a way to make amends when they've fallen short.

Beyond Catholic confession, which is sometimes called the Sacrament of Reconciliation, other Christian sects offer a more direct line between the believer and God: prayer. Say what you've done wrong and ask for forgiveness.

In Judaism, Yom Kippur is the holy Day of Atonement, which features self-inflicted chest pounding, done while enumerating sins and mistakes committed both "under duress and willingly." Another part of the holiday involves making rounds, apologizing to those you have wronged. Because Yom Kippur falls near the Jewish New Year, there is a sense of working toward a fresh start, with God and your community. Or as my brother Sam used to say when he was small, after he had misbehaved and subsequently apologized, "New day! Can we please have a new day?"

For Hindus there is Prāyaścitta, which translates to some combination of atonement, penance, and expiation. Depending on the sin that has been committed and who the sinner is, the antidote might be public or private. The recipe for absolution can range from taking on the role of beggar, to committing to celibacy, to making a charitable donation, dieting, or embarking on a holy pilgrimage.

In Islam it's called Istighfar and it takes place at dawn

each day and on Thursday nights. It requires the repetition of the phrase *astaghfirullah*, "I seek forgiveness from Allah." Some Buddhists, in accordance with the suggestion of a sixth-century monk called Zhiyi, practice a specialized meditation in order to attain visions of karmic redemption.

According to the Swedish anthropologist Åke Hultkrantz, who devoted his life to studying the indigenous people of the Americas, the Inuit, Athabascan, Ge, and Tupi tribes believed that physical disease came from failing to obey social norms, and that "disease is often abolished after the patient's 'confession' of the taboo offense to the medicine man." Maybe this was a way of describing how guilt eats away at you.

The people of Central and South America were ritualizing confession long before the conquistadors arrived with Catholicism. As Hultkrantz writes, "Confession was a traditional institution among the Maya as among the Inca Indians . . . The similarities of Catholic and Maya beliefs led then to the conclusion that the apostle Thomas must have preached the gospel to the Maya Indians." How many lives would have been spared if the Spaniards could have managed to see what they shared with those they were conquering?

Ritualized atonement hasn't always been about things an individual has or hasn't done. The historian of religion Karen Armstrong writes of how the kings of the Zhou dynasty (who ruled parts of eastern China around three thousand years ago) were held personally responsible for the changing of the seasons, natural disasters, and the weather. They were politically powerful but shackled by strict ritual meant to keep their world turning. The palace became a kind of control room for the cycles of the year, where these kings attempted to pilot nature's transformations. To cajole spring to bloom, the king stood in the northeast corner, wearing green, eating sour food. For summer, red. For fall, white. For winter, black. It was his sacred duty to will the seasons into reality and on schedule. When they didn't arrive on time, or when the rains didn't come or the Earth quaked, when the people were hungry, the king issued a public apology, professing that his failures as a leader were the source of nature's unreliability. He would then make a sacrifice at the "Earth altar." Imagine his anxiety as he watched the skies for signs of what nature had in store.

But you don't have to be a major faith or empire to ritualize confession. Alcoholics Anonymous may or may not be a religion, depending on your definition. (It's certainly a theistic institution with dogmas, customs, ritual, expressions, culture, and a single authoritative text.) Recovering alcoholics are expected to enumerate their own role in any difficulties they've faced. When they reach step five of twelve, they read the list aloud to God, before a human witness.

Later, at step nine, they must accept responsibility for previous transgressions and make amends. In other words, they must atone.

Same goes for therapy. Not exactly a religion, but it still involves customs and rituals. It's meant to be a pathway to understanding why we are the way we are. It's made up of different sects and denominations that agree on basics but disagree on details. And it requires unburdening oneself to a confessor who, like a priest, has taken a solemn oath not to divulge your secrets to anyone else.

If you commit a crime and then cop to it, the state also ritualizes your confession, sometimes by having you allocute before the court. Sometimes you have to write your misdeeds out and sign. Either way it's government-regulated confession. Serving your sentence might be seen as your atonement.

After my father died, there were a series of biographies, profiles, and documentaries about him. Some were accurate and fair, capturing who he really was. Others had factual errors or were written by people who just didn't get him. Sometime in that first year or two of our new grief, my mom, not serious, turned to me and said, "Maybe biographies should be illegal. You can always trust autobiographies because we humans always tell on ourselves by accident." She was joking about the first part, but the second has stayed with me all these years.

The writer Ta-Nehisi Coates tells of how, when he got in trouble with his teachers as a kid, his mother would have him write about what he had done. He would answer a series of questions that led him to examine his own behavior, exploring how he felt and what he thought about the situation. In his stirring open letter to his son, *Between the World and Me*, he writes that "these were the earliest acts of interrogation, of drawing myself into consciousness." For those of us who have no single sacred guidebook that outlines the dos and don'ts of life, this kind of self-examination is required. We must each suss out our own canons from what our conscience and experiences teach us.

Of all these groups, Maya and Inca, Christians and Muslims, alcoholics and writers, the people who should be most willing to regularly ritualize their apologies for having been wrong are those of us whose philosophies are rooted in science.

There is nothing more central to science than error correction. Scientists are not infallible. Quite the opposite. The greatest minds in history have often been wrong about lots of stuff. But the defining difference between science and religion is that you're a better scientist if you take the ideas of the people who came before you, the people whose shoulders you stand on, the people who taught you everything you know—your teachers, your heroes, your mentors— and disprove them. Then you've done your job. Doing the same does not make you a better pastor, rabbi, cleric, or monk; upholding tradition does.

My dad once wrote, "In science it often happens that scientists say, 'You know that's a really good argument; my position is mistaken,' and then they would actually change their minds and you never hear that old view from them again. They really do it. It doesn't happen as often as it should, because scientists are human and change is sometimes painful."

It's true. These words: *I was wrong. I made a mistake. I did something bad. Something selfish. Something mean. Something stupid. Something thoughtless. I'm sorry.* Why is it so hard to say them? Especially since they're true for every single one of us?

One evening, after the birth of our daughter, I thought I should lead by example. I closed my computer and went upstairs to apologize to Jon. I apologized for falling asleep while he was talking about work the night before. I apologized for spending more than we had agreed on something. I apologized for giving him a hard time about folding up a piece of paper I felt was too important to be folded up. It didn't matter that it was folded. I was just being controlling and neurotic. And you know what he said? "That's okay. I'm sorry I folded it up." And then we kissed.

It was a different kind of sacrament of reconciliation. It was small, somewhat spontaneous, but definitely a ritual. And it gave me a little clarity on why we really need them.

At the beginning of the last century a German-Dutch-French anthropologist called Arnold van Gennep tried to define *ritual*. His work is, in ways, predictably dated. He used the language of a person brought up to believe his continent is the best, that the people who live there and look like him are intrinsically superior. Were he alive today he would probably have to do some atoning. But he did manage to coin a great phrase: *the rites of passage*. His book by that name explores a wide variety of rituals, emphasizing what they share. He saw rite, or ritual, as a kind of three-step path from a state of separation to a state of togetherness. Things are one way: separate, apart. Then there is a transformation, a transition, a change. When that is complete, everything is together, united or reunited. For atonement-related rituals it would go something like this:

You are separated by your sins or shortcomings from the way you believe God or your community wants you to be, from your potential, from some kind of purity. You confess, do penance, apologize, and make up for what you've done. And thus you are reconciled with God, or godliness, or your community's approval, or your ability to sleep at night.

Anthropologists still pretty much look at rituals this way.

This middle step is sometimes called a "liminal phase," and, in turn, the before and after are sometimes called the pre-liminal and post-liminal phases. *Liminal* comes from the Latin word for threshold. So ritual is a portal into another world. This is just as true for rituals that happen once a year—say, decorating the Christmas tree—as those that happen once in a lifetime—like a funeral.

However, there is no innate astronomical or seasonal cue for us to atone. We should probably do it often, whenever we have the gnawing sense we've wronged someone. And yet, as is usually the case, having a specified time makes us more inclined to do something. It's harder to skip a commitment when you've set the moment aside ahead of time.

My mother and I have often thought that March 4th would be a good date for such a secular holiday. When you say it out loud, in English, it sounds like a bold command. It's a pun that seems to cry out a directive to improve or "evolve."

In a footnote to the book *Pale Blue Dot*, my dad describes the events of March 4, 1953 BCE, when the five planets visible to the naked eye aligned perfectly, "strung out like jewels on a necklace near the great square in the constellation Pegasus." This was observable in China and may have been "the starting point for the planetary cycles of the ancient Chinese astronomers." Up until that moment they had a completely different view of the universe and our place in it, but faced with new evidence, they changed their worldview, abandoning what no longer made sense.

My brain cannot help drawing connections between the verb *march* and the month March, the adverb *forth* and numerical term *fourth*. It's nothing more than a happy coincidence, but if it hadn't been this event it would have been another. March 4th is as good a day as any to repent. This needn't be a day to beat ourselves up for being human, but instead to shake off what can no longer stand up to scrutiny, and to bravely *march forth* into the personal and philosophical unknown.

My mom often references an essay by my dad where he advocated for acknowledging our mistakes. In it he enumerated times he had been wrong. Like writing critically about *The Simpsons* while not having really watched the show. He realized later he'd failed to understand that it was subversive, not a glorification of stupidity. And there was the time when he went on TV, exhausted, right after Sam was born in 1991, and said that the oil fires in Kuwait would have a dramatic, immediate effect on the Earth's atmosphere. Well, they didn't. My mom talked about this piece so confidently and with such detail that I started to think I had actually read it. I told other people about it, not mentioning that I had not ever laid eyes on it—a lie of omission. When I went to look it up, Google was no help. I asked my mother. She couldn't find it either. It turns out it was never published. The editor had rejected it because he felt it would be too damaging to my dad's reputation. When she realized her mistake my mom texted me "so sorry" with pink and red hearts. That was as simple and perfect a sacrament of reconciliation as I could imagine.

I became determined to find some trace of the essay, maybe just a few sentences outlining the gist. I tried to track it down in the Library of Congress. Maybe I'd get to see part of it in my dad's handwriting, the fragment of an idea that never quite saw the light of day. At my request, a librarian there very kindly went through a box of miscellaneous materials from 1982 to 1996, coincidentally the exact period of time I had with my dad, the years between my birth and his death. But there was no evidence of the essay in there. Maybe it had been lost, maybe it only ever existed as a conversation, just sound waves in the air. I'll likely never know. But I couldn't stop thinking about the box where the essay had not been. I imagined the librarian going through the dusty old things that used to live in the house where I grew up. Things my dad had held. Papers I had probably seen when coming to his study, hoping to distract him from his work. I could picture all the familiar lined yellow legal pads and brown accordion folders, stacked among the books. His sometimes nearly illegible cursive, sometimes perfect printed block lettering, in the ink from the "extra-fine" Flair pens he always used. How ordinary they seemed to me then, and how sacred they seem now. How sorry I am I didn't cherish every minute, every tiny detail, conversation, inside joke, and quiet moment. But that's the thing about death—it makes you appreciate life. It's almost impossible to appreciate something without facing its absence. Just as we cannot improve ourselves if we cannot acknowledge where we've floundered, and atone.

chapter six

Coming of Age

By convention there is color, by convention sweetness, by convention bitterness; in reality there are atoms and space.
— DEMOCRITUS

How often when we're small do we hear some relative who lives a great distance away become bewildered upon seeing how much we've grown? I remember so clearly thinking to myself, *This is ridiculous, of course. I'm ten now. If you saw me last when I was three and that was seven years ago, what else could I possibly be?* And yet, here they were, intelligent, worldly people aghast at simple arithmetic.

It took me many years to understand, until I was old enough to have friends with babies whom I saw sporadically, babies who grew into children and then adolescents while I wasn't looking. And now here I am, left stunned at time's power to make, for example, a little girl into a young woman. It feels like the result of a magic spell, something supernatural.

My mother has always said there's a problem with the word *supernatural*. Literally, it means "above or beyond nature." But most of what we categorize as supernatural—for example, witches, monsters, or ghosts—is either less compelling than nature or really just an extension of nature itself, often a metaphor for it. We use the word to describe a kind of magic, some unexplained phenomenon or idea that gives us chills. Our lack of understanding seems to imbue it with power. But I find much more chilling the power in the things that we do understand.

Take the legend of the werewolf. Long before Michael J. Fox was howling at the moon in a varsity jacket, the idea of a person turning into a wolf had appeared in folklore from the nomadic empires of central Asia all the way to pre-Christian western Europe. The first recorded mention may be from as early as 2100 BCE, in the *Epic of Gilgamesh*, depending on which scholars you agree with. By the first century CE it had become part of the Roman zeitgeist. Petronius, a courtier of Nero's, writes of a soldier having his wounds dressed in his novel *Satyricon*. A character exclaims, "I saw at once he was a werewolf and I could never afterwards eat bread with him, no!"

Sometimes werewolves were akin to warlocks, and sometimes they were the victims of a curse, something that took them over without their consent. But in some form they were there, shape-shifting and reappearing, century after century. Why? What made this idea stay with our species for two millennia?

There are actually two very rare but real medical conditions that reflect this myth. One, hypertrichosis, sometimes called werewolf syndrome, causes hair to grow in all sorts of places it normally does not reach. The other is a psychological condition in which you think you are an animal other than a *Homo sapiens*. It doesn't have to be a wolf—it could be a tiger or a duck—but the name of the condition, *clinical lycanthropy*, is Latin for "wolf human." Of course, hypertrichosis and clinical lycanthropy are extremely unusual problems. Not really relatable enough to keep a legend going since ancient times.

So what is this story really about? What makes it resonate?

Imagine it for yourself. There you are, minding your own business, when something takes hold of you. You change unwillingly, both physically and psychologically. You transform into something hungry, wild, and full of mysterious, dangerous urges. You sprout hair in unexpected places. You lose control of yourself at night. This may be the stuff of nightmares, but there is nothing more natural than puberty, the real-life experience that I suspect is at the heart of this myth. The narrative of the human who suddenly changes into something else is not beyond nature, it's part of it.

And for our species, it's crucial. The transition from being a child to being able to make one is so important that it has been cause for some of the most elaborate and imaginative rituals the creatures of Earth have ever devised.

The Amish break their strict code of conduct to let their teenagers explore the outside world during *rumspringa*. The Apache bless their daughters with pollen before a four-day ceremony, during which dances are performed at dawn each day. Teenage Maasai enter manhood with a dance party and a circumcision. In Bali there is ritual dental work, in particular a slight filing down of the canines, which represent lust and other grown-up sins. In Ethiopia there's cow jumping. The young men of the Sateré-Mawé tribe of the Amazon insert their hands into gloves filled with angry, venomous ants, and as they're being stung, they dance, proving their manhood. For some ancient Greek boys there was an animal sacrifice and a haircut. Confucian tradition also gave boys a grown-up hairdo and a new name at twenty. By the twelfth century, Japanese adolescents were receiving a ceremonial makeover called *genpuku*; for boys this featured their first sword and coat of armor, while girls blackened their teeth, applied makeup, shaved their eyebrows, and—as in Greece and China—changed their hairstyle. The modern incarnation of *genpuku* is the Japanese holiday of *seijin shiki*, or Coming of Age Day, a national celebration of everyone who has reached legal adulthood in the past year.

Among the Mormons the coming-of-age ritual is an adventure: the mission. After high school, teen members of the Church of Jesus Christ of Latter-day Saints are obliged to leave their families to proselytize for two years. Until recently it was only for boys, but now when we fly into Salt Lake City on our way to visit Jon's sister, we see not only name-tagged boys in suits but also girls in shin-length skirts, ready to leave the nest.

The Baha'i, members of a religion founded in the nineteenth century, believe that the Gods and prophets of all the modern monotheistic religions are reflections of the same entity. No one is automatically Baha'i because their parents are. At age fifteen one is offered the choice to join the faith. While their coming-of-age ritual doesn't have the same pomp and circumstance of so many others—it includes filling out a registration card and sending it in—it's still a rite of passage.

For Jewish adolescents there are bar mitzvahs (for boys) and bat mitzvahs (for girls), which usually consist of a morning of reading ancient texts before your congregation and an evening of dancing to modern songs with your friends.

These gateways into adulthood get dressed up differently by societal norms and cultures, but deep down each one is about the release of gonadotropin into the pituitary gland, about estrogen and testosterone, and the new possibilities of sex, pregnancy, parenthood, and responsibility. That is what we're really celebrating at every *quinceañera* and bar mitzvah, every Roman Catholic confirmation, every cotillion, every debutante ball, every sweet sixteen. This is about biology, sexual maturity, and the survival of the species. These parties are the community's wish that their DNA will live on after they are dead.

Jon's family has historically been members of a variety of Christian sects. His father attended Catholic school but no longer practices. His mother is also now secular but is the descendant of all manner of Protestants, from Puritans to Unitarians. For ten years, when Jon was a kid, she was married to a local minister of a First Congregational Church. As Jon describes them now, his former stepfather's sermons presented the Bible as more of a metaphor. His advice was about how to be kind and thoughtful, not fanatical. Once, Jon wrote *hi mom* in huge letters in the snow of their front yard, and his stepdad used this example to tell his congregation that spontaneity and romance need not be between only partners but present in any relationship with someone you love.

His Sunday school teacher, however, was more of a literalist. While Jon was a very good middle-school student all week, he could not bring himself to behave in Sunday school. He couldn't understand why this old book, the Bible, was so much more important than all the others. Jon's mother told him that once he was confirmed at thirteen he could decide for himself about going to church, but he would have to prepare a personal statement to read in front of the other Sunday schoolers. This was his coming-of-age ritual. He completed it, and from that day forth, he had his Sundays to himself.

Years later, he told me that when he pictured God in his mind's eye as a child, he imagined "Santa Claus without the suit." As he got older, the personification and behavioral decrees fell away. Slowly, God became a word for the way the universe works, something like Einstein's God of Spinoza. By college he called this physics. He had no epiphany, no rebellion, no crisis of faith. His belief faded as he learned more. For him, the true ritual transition to manhood wasn't giving the speech before the class but being allowed to decide what he really believed.

Certain rituals seem to perform the rite of passage more explicitly than others. A thousand miles off the eastern coast of Australia, on the Pacific island of Vanuatu, boys approaching manhood climb to the top of wooden towers up to ten stories high to greet their friends and neighbors from above, take in the view, sing out, tie two vines to their ankles, and jump off. The vines act as a kind of organic bungee cord, and the young men are almost always fine. In fact, Ni-Vanuatu (that's the demonym) guys continue land diving regularly over the course of their lives, reenacting an ancient myth about an escape from an abusive marriage. But the maiden voyage is special: at the end of your first dive, your community rallies around you, and your mother destroys your favorite childhood object. Could there be a more perfect embodiment of puberty?

Puberty, of course, takes place at different ages for different people. For boys there isn't a singular physical event that marks their entrance into manhood. Maybe you have a faint mustache at nine, but your voice doesn't crack until fifteen. What makes you a man? Young Yupik men in southwestern Alaska receive a tattoo of a symbol called *ellam iinga*, or "the eye of awareness," upon their first kill as a hunter. Young Yupik women are not expected to hunt, so they get their tattoo at a different occasion: their first period.

I was twelve when I got mine. I came home from middle school and thought something was amiss. I ran to the bathroom. Sure enough. Health class had prepared me for this, and I felt relieved that things were moving along on schedule. But what I was not prepared for—and still don't fully understand—is that when I emerged from the bathroom, my mother took one look at me and said, "You got your period." How did she know? From the expression on my face, I guess. But I was astonished. It felt, for lack of a better word, supernatural. Like she had superpowers—which she did—and now I did, too. I could, as she had, make a person. (A power I carefully refrained from using for another twenty-two years.) My mother was proud of me, excited for my entrance into womanhood, messy and crampy though it might be. She congratulated me as though something lucky had happened. She took me in her arms and made me feel that this was cause for celebration. Despite the fact that, thirty-some years earlier, when she had gotten her first period on her thirteenth birthday, she was greeted with a slap in the face. A literal one. From her mother. This, unfortunately, is a Jewish custom. Like so many Jewish customs, the *why* is a gray smorgasbord of metaphors: it's to knock some sense into you, to awaken you from the slumber of childhood, to make your cheeks rosy so boys will like you. When my mother told me this happened to her and I asked her why Grandma would do that, why we would have such a hostile custom, she told me the subtext first, as she often does: "It's to punish you for being a woman."

Looking back, my true "coming of age" was still years away. When I was in my late teens, I accompanied my then boyfriend to visit the family home of a friend in another country, where we took psychedelic mushrooms. Hallucinogens are featured in many coming-of-age rituals in societies around the world, but those are carefully planned traditions. This was more, shall we say, impromptu.

At first it was really great. My boyfriend (let's call him Milton—not his real name, needless to say), his really cool friends, and I spent most of the afternoon outside, doing what can only be described as frolicking. I wasn't hallucinating yet, but I do remember quite clearly thinking that the bright blue wind pants I was wearing were just about the most exquisite pair of pants known to man. I felt happy.

Later, as the sun was setting, I was alone on a balcony of the house where we were staying, smoking a cigarette. A sense of impending doom washed over me. In the distance, I saw red lightning strike, clear as can be. My first hallucination.

Things went downhill from there, and quickly. I thought I could see the Earth from space. It was tiny, alone, meaningless, and inconsequential. I was in the throes of a full-blown existential crisis. All the painful concepts I was already aware of but had kept isolated from my feelings were suddenly viscerally real: Nothing lasts forever. I'm never going to see my father again. Everyone I love is going to die. I'm going to die. I'm tiny in a very big universe. The sun will burn out in just a few billion years. No matter what I do, in the grand scheme of things, nothing matters.

None of this was news to me. I'd just been very good at pushing these ideas out when they elicited a sense of palpable, overwhelming sadness or panic. But with the drugs in my system, I lost control, and all the emotions I had been keeping at bay flooded in. Soon I couldn't understand why anyone would refrain from killing themselves. I just couldn't wrap my mind around a single thing to live for. The only thing I could think to do was call my mom.

"Can we please not call your mom while we're on mush-rooms?" Milton begged. "It's too awkward."

"But do you think she's ever felt like this?" I asked.

Milton assured me that, since my mother had been a very hip person during the 1960s, it was a virtual certainty that she had.

We didn't call her. Instead, I tried hiding from my own thoughts under a blanket, which, no surprise, proved fruit-less. I couldn't get away from the horrifying finiteness of it all. I spent the night weeping about the terrible extinctions to come, until finally the effects wore off, and sometime in the wee hours of the morning I was able to fall asleep.

On the journey home, now sober, I was still despondent. Like a billion teenagers before me, all I could think was *What's, like, even the point?* I was blue. And I was blue for a while. I thought I had permanently lost a kind of joie de vivre that would never return, that from now on, whenever something good happened there would be a nagging voice saying, "Yeah, sure, it's great you got an A, but the sun is go-ing to implode, so . . ." I thought I had unleashed a kind of eternal buzzkill, something I'd never be able to put back in the bottle, ruining every future pleasure.

Many weeks, maybe months, after I returned home, my ennui slowly dissipated and another kind of perspective started to emerge. Not a new one, exactly, but one that my parents had instilled in me long before. I already *knew* that just being alive at all is astonishing and beautiful, but I don't know if I had ever really *felt* it fully before. I started telling myself that no matter what tomorrow brought, each little moment on Earth was still meaningful. And that if life went on forever it would not be as precious. I started reminding myself that even though I will certainly die someday, I am alive right now, which is an incredibly lucky thing. Slowly, this idea started to bring me a kind of giddy joy, almost like butterflies in my stomach. It took a while, but eventually I was happier than I had been before the trip. And I was happy because of a deep understanding of the finite nature of life, not in spite of it. This, to me, felt like adulthood. I don't believe ignorance is bliss. I think understanding is bliss, but, to get to the joyful part, sometimes you have to face the terror head-on. Once I could admit to myself how truly tiny we are, how short our time is, and still love life, I felt like a woman.

There is something about facing fear that defines growing up. Doing something hard, freeing yourself, taking your fate into your own hands—these are the portals into adulthood across the planet. Even if it's just a matter of getting your driver's license.

Right now, my daughter is transforming from a baby to a kid almost imperceptibly each day. Someday she will make another transition: from kid to adult. It will also be a long series of small steps. We'll throw her a party if she wants one. We'll help her prepare for a ritual or ceremony of her choosing. But the moment she feels that change, the moment she crosses that threshold into womanhood, will take place in her mind, not necessarily in front of her community. And its significance may only reveal itself in retrospect.

For better or for worse, it was an impromptu vision quest that (eventually) led me to a place where I was no longer burying my deepest fears about the nature of existence. It freed me from a kind of cognitive dissonance between what I knew and what I felt. It allowed me to stop pushing information out of my head when it made me uncomfortable. Or at least, to do so a little less. For me, this was a major step toward growing up. I had to give up some of my illusions, which was painful. But what I gained—a deeper sense of what was real—was worth it. For our species to survive we all must, in our own ways, jump off the proverbial tower, let our security blanket be destroyed, and go forth into the terrifying wonder of life.

*

chapter seven

Summer

A single sunbeam is enough to drive away many shadows.
—*St. Francis of Assisi*

The Earth would die
If the sun stopped kissing her.
—*Hafiz*

When I was little, Maruja would go home to Lima around Christmastime to visit her sisters. She told me that, while home, they would go to the beach, because in Peru it was summer then. This was absolutely unbelievable to me. As in *not believable*. I thought it was a trick, a joke, maybe a riddle of some sort. At some point my dad explained the axial tilt of the Earth, the equator, and the hemispheres to me. He explained that the longest night of the year in Ithaca—the winter solstice—was the shortest night of the year—the summer solstice—in Lima. I had a lot of follow-up questions that, lucky for me, my dad was happy to address in detail. He was able to provide supporting evidence, and slowly I came to accept this, despite the fact that it didn't *feel* true. While I watched the snowfall, I tried to imagine people in the southern hemisphere basking in the heat. In the cold the sunshine felt very far away.

That feeling of sunshine is a true mood enhancer. The ultraviolet radiation we get from the sun releases endorphins in our brains. It's a real chemical reaction, a scientific connection between our bodies and our closest star. How beautiful is that? How astonishing that being bathed in rays of light from a 4.6-billion-year-old mass of hydrogen and helium located 93 million miles away can make us feel happy?

Until recently we humans believed that life required sunlight. That's not absolutely true; there are creatures in the depths of the ocean that have no access to light. And who knows what life might require on other worlds. But most Earthlings are very dependent on the sun—for example, those of us who happen to be mammals who eat plants or eat other animals who eat plants. Without the sun, photosynthesis would not be possible, and without plant life this would be a barren rock.

We owe the sun our lives.

If worship is, at least in part, about gratitude, about bowing down to the source of our blessings and bounties, then our bright, hot neighbor fits the bill perfectly. And for this reason, a vast array of human belief systems have featured a sun god.

Take Helios, familiar from the cover of my Greek mythology book. His name still means "sun" to us in words like *heliocentric*. After a few centuries he popped up again in ancient Rome, under the pseudonym Sol, eventually giving us more sun-related words, including *solstice* and *solar*.

Ra, the ancient Egyptian sun god with a falcon's head, might be the only other sun god I learned about in school. The ancient Egyptians worshipped the sun as if each daybreak was like the second coming. Solar cults were so influential in one city that when the Greeks got there, they called it *Heliopolis*, or Sun Town. Ra wasn't even the only sun god in Egypt. There were gods for sunrise and for sunset, gods for searing summer heat, and gods for where the sun might go at night.

The ancient Babylonians, who lived in what is now Iraq, worshipped Utu, a sun god who transformed at night. When the sun set he became the god of the underworld. A millennium earlier, the Sumerians also worshipped him, calling him Shamash.

In the Shinto religion, the traditional belief system of Japan, the divine forces called *kami* include one's ancestors, elements of nature—like trees and thunder—and even abstract concepts like growth. But nothing gets higher billing than Amaterasu Ōmikami, the sun goddess. According to the Shinto philosophy, the current royal family of Japan are her descendants. A mirror, the reflector of her light, is one of her three sacred symbols.

According to the Harvard Divinity School professor Jacob K. Olupona, in the highlands of Ethiopia, the Kafa, Seka, and Bosa peoples see their human king as a kind of sun god, and because of this he is forbidden to eat during the day, to prevent two suns from shining at the same time.

For the Inuit, whose northerliness hides the sun for long periods, the sun goddess Malina is being forever chased away by her brother and stalker, the moon god. His obsession with her causes him to forget to eat, and as he wastes away the moon wanes. After he finally goes to eat, his return heralds the new moon. A solar eclipse is the moment he catches her. Throughout the Americas, from the Algonquin of Canada to the Araucanians of Chile, people have deified this yellow star. In the heyday of the Incan empire, Inti was worshipped as the incarnation of the sun and the brother and husband to Mama Quilla, the moon goddess.

This is a tiny sampling of our many solar deities. There are countless more documented by historians from every inhabited corner of this rock, and likely many more that have been lost to time. Even Jesus called himself "the light."

Light itself seems to be almost universally sacred. (I hesitate to use the word *universally* about anything because, as my dad taught me, we only have information about Earth, nowhere near the whole universe.) How often is light used as a metaphor for hope or for remembrance? How many rituals center around a candle, that small, short-lived sun of human invention? Is not every birthday candle, menorah, Chinese New Year lantern, firework, Christmas light, and Olympic torch a little homage to the sun?

Over the generations, we humans have spent a lot of time looking at our star. Even now, with access to so many interesting things to stare at, nothing really compares to watching the day end with a wildly colorful sunset. We've devoted a lot of time, energy, and resources to building new ways to view our star from the surface of our planet. Special effort has gone into architectural observatories all over the world that appear to have been built for ritual viewing of the sun during the summer and winter solstices, places to meditate on these extremes of day and night.

There are very famous ones like Stonehenge—built by the ancient peoples of the British Isles—or Angkor Wat in Cambodia, or the Temple of the Sun, built by the Incas at Machu Picchu. Lesser-known ones like Pueblo Bonito in Chaco Canyon and the spiral Sun Dagger on Fajada Butte built by the Anasazi a thousand years ago in what is now New Mexico, and the megalithic observatory excavated in 2005 near Calçoene, Brazil, and the grand Incan calendar at Cuzco that impressed the Spanish so much that they felt it had to be destroyed in the sixteenth century, so as to prevent the indigenous population from knowing when to celebrate their "heathen" festivals. They kept the foundation and built a cathedral on top of it, not realizing the irony that Christmas and Easter were built atop the ruins of pre-Christian winter solstice and spring equinox celebrations.

The celebrations and rituals that took place in these tailor-made sun churches are lost. We have some remains of the set, but the scripts, the costumes, the faces of the actors are gone. What I wouldn't give to be able to witness those summer solstice festivals. At Machu Picchu and Cuzco they would have taken place in what we call December, summer for the southern hemisphere. It's confusing because we live in the world created by invaders. And it shows. Throughout this book, I try to compensate for my own hemispherical bias, but I must confess that when I picture the summer solstice I think of June on the calendar. About 90 percent of humans do live in the northern hemisphere, but not all of us. If the Inca had shown up on the shores of Iberia convinced they had a god-given right to take over, maybe the Spanish would be celebrating Quyllurit'i, the Incan pre-winter solstice festival, during their summer.

No matter how hard we revere it, though, the sun will eventually die. The life spans of stars are measured in billions of years, but they do eventually meet their end, like all of us. But when they do, we don't get the news for a little while. The light we see from stars is old. This is because light does not travel instantaneously. It's very fast but not instant. To quote from the book version of *Cosmos*, "If you are looking at a friend three meters (ten feet) away, at the other end of the room, you are not seeing her as she is 'now'; but rather as she 'was' a hundred millionth of a second ago." The farther away the star, the longer it takes for its light to get to us. Some of the stars we can see at night are already dead. Starlight is a kind of time travel. A vision of the past.

When I was little my dad and I spent a lot of time talking about time travel. And a lot of time watching, re-watching, and analyzing the *Back to the Future* trilogy. I was already obsessed with understanding what it would be like to live in another era.

"We are time traveling," my dad would say, "one second at a time into the future!"

This was true but not what I meant. I wanted to explore other centuries. Because of this, my singular obsession on trips to Disney World was Spaceship Earth, the enormous golf ball that serves as the emblem for Epcot. I would beg, negotiate, and manipulate my parents into taking me on the Walter Cronkite–narrated "history of communication" over and over. I could not tire of cave paintings, animatronic Phoenicians trading scrolls, or the smell of Rome burning while messengers raced to spread the news. Along with one afternoon spent in Colonial Williamsburg, this was about as close as I could get to time travel. Wanting more, I devised other, less literal methods.

Sometimes, especially on the beach, I like to play a little game. An old forest or the foot of a mountain will work, too. But I like the ocean. Looking out to sea on a summer's day, I create a little window with my hands, framing a view that could have been there a hundred years ago, or maybe a thousand, and I imagine the lives of the people who looked at this view through the ages. I try to imagine what they might have thought and felt. I try to time travel into their minds. Surely, no matter who they were or when they lived, being enveloped by the rays of our nearest star they felt something. On the days that we call the summer solstice, it's safe to assume some version of *I can't believe how late it's light today* or just *Today was really long* crossed their minds, like it crosses ours.

I still play that game, but now I mostly long to time travel to sometime between November of 1934 and December of 1996, the period during which my father was alive.

I have a few ways I can do that. I can time travel with the help of my memories of him, visiting him in my mind, although memories are fallible. As the years go by I am less clear on details, less sure I remember everything about him exactly right. Luckily I have other ways, too. Mostly collaborating with my mom, he wrote about two dozen books and countless essays. I can still read words like these:

> What an astonishing thing a book is. It's a flat object made from a tree with flexible parts on which are imprinted lots of funny dark squiggles. But one glance at it and you're inside the mind of another person, maybe somebody dead for thousands of years. Across the millennia, an author is speaking clearly and silently inside your head, directly to you. Writing is perhaps the greatest of human inventions, binding together people who never knew each other, citizens of distant epochs. Books break the shackles of time. A book is proof that humans are capable of working magic.

Now that he's gone, those very words perform that magic trick. And because I knew him I can even hear them in my head in his voice. It's like basking in the light of a dead star.

For most of history, if you loved someone and they died, that was the only way you could ever hear their voice again: in your head. Or maybe if you were lucky they would have some relative whose voice sounded like theirs, similar at least. But just recently, our species had managed to figure out a way to record sound. Now we can do it very easily and as often as we want. This has been a big change over even my lifetime.

I still have the same best friends I had growing up in

Ithaca. We are a very close-knit bunch. Perhaps, in part, because we lost one member of our group the summer after our first year at college. It was a car accident, followed by weeks in the hospital, before Brent died, not long after the summer solstice. Teenagers nowadays have hours of video of their friends, thanks to the wonders of science and technology. We didn't have that then. After he was gone we would call his cellphone, a flip phone, to listen to his voicemail greeting, just to hear his voice again.

Even though my dad died five years before Brent, his case was different because of the unusual nature of his work. He had been on TV a lot. I had VHS tapes, then later DVDs, with thirteen hours of my dad hosting *Cosmos*. There he is, saying words that he and my mother wrote (together with the astrophysicist Steven Soter), explaining what he knew about our place in the vast majesty of the universe.

Even later, I could Google him and watch videos I had never seen before of him as a guest on the *Tonight Show*. Or him being interviewed on the street holding me as a baby, something I had not known existed. His light was still reaching me, through technology, through science, the summer of his life still warming me during the endless winter of his death.

As I watched and read, I had the sense that there was still somehow more of him in my future. The more footage there was to uncover, the less he only existed in the past. It helped. I think of how much harder it must have been to lose your parents when you didn't have any way of seeing them again. I felt very lucky.

In Karen Armstrong's *The Great Transformation*, she wrote that, during the seventh century BCE, when Chinese culture centered on honoring one's ancestors, "the son revered his father as a future ancestor." After the death of a patriarch, the son would enter a kind of mock death in solidarity; sleeping apart from his family, closer to the earth and the elements, like a corpse, almost provoking his own health to fail while, he believed, his father transitioned into the universe of their ancestors. At the ceremony that marked the end of the mourning period, and the dead father's successful transition, the dead man's grandson, the son of the son, performed a kind of impression, taking on his mannerisms, his voice, his laugh, all the things lost to death. On this occasion the father would bow to his son as though he were his late father. How haunting it must have been sometimes. How unconvincing at others. But in a world without photography, without audio recording, it was one last chance to see your dad.

There is one more way I time travel back to my father. When I was little he told me that air particles stay in our atmosphere for such a long time that we breathe the same air as the people who lived thousands of years ago. I think about that often now. I can take a deep breath and know that some fraction of those particles were once breathed by my dad. What an intimate thing it is to breathe the air of someone you loved.

In his book *Caesar's Last Breath: Decoding the Secrets of the Air Around Us*, Sam Kean writes: "All the world's roads and all the world's canals and all the world's airports in the history of humankind haven't handled nearly as much traffic as our lungs do every second. From this perspective Caesar's last breath seems innumerable, and it seems inevitable that you'd inhale at least a few molecules of it in your next breath."

And it works the other way around as well. The air you're breathing right now, this second, involuntarily, automatically, it's not just the old air of Jesus and Muhammad and Cleopatra, but also the new air of future generations. And not just people. If we humans manage not to destroy our planet completely, this air, our air, might be breathed by creatures not yet evolved. New beings we can't even imagine yet. We are, after all, someone's distant future and someone else's ancient past.

Someday when our daughter is bigger, we'll go somewhere in the dead of summer, maybe on the solstice itself, somewhere outdoors, somewhere beautiful and ancient. Or just look straight up on a starry night. We'll find a way to block everything modern and new from our view. We'll make a little tunnel with our hands and try to imagine what it would have been like to be among the first humans to set eyes on this view. We'll imagine what our world was like when the light we now see left those distant stars eons ago. Was there life here yet? Anyone to see some other configuration of stars? We'll time travel together, inhaling those same air particles they might have once pulled into their lungs, and release them back into the world.

*

chapter eight

Independence Days

*I love America more than any other country in this world, and,
exactly for this reason, I insist on the right to criticize her per-
petually.* —JAMES BALDWIN

The last place the colonizer leaves is your mind.
—HARI KONDABOLU

I did once meet someone who was almost like a time trav-
eler. I used to volunteer at a center for "at-risk" teens in
New York, helping them with their homework, mostly social studies. Some of the kids were homeless, some were vi-
ciously bullied at school because they were gay or trans or
otherwise didn't fit into the binary gender norms of years
gone by, and many were new immigrants to the United
States. I would often start history lessons with what was
usually perceived to be a trick question:

"What language are we speaking?"

Tentatively, unsure, the kids would usually answer, "English?"

It wasn't a trick question.

"Yes!" I would reply.

The student would be relieved to discover I had not lost all grasp of reality.

"But why?" I'd ask next. "We're not in England. Why do we speak English?"

Kids from francophone African countries always knew the right answer. Most French African nations won their independence in the late 1950s and early 1960s. Those fights for freedom in places like Côte d'Ivoire, Mali, and Niger are still fresh. Those kids knew the veterans of these political revolutions. There was no way for them to not know why they spoke French. And it was easy enough to extrapolate why we spoke English in America.

One day, a young guy who I took to be in high school, or maybe a late-blooming college student, walked in. I thought he must be a new tutor, not because he seemed too old to be a student, but because he was in the black pants, white shirt, yarmulke, *peyos*, and tzitzit of an Orthodox Jew, the uniform of my forefathers. My own instant, ancient chauvinism, the type I didn't even realize was there, told me, *He is like you* and therefore not in need of help. This bubble popped the second he opened his mouth.

"I need a tutor!" he announced in a Yiddish accent, as though it was an emergency.

"Okay. In what subject?" I asked.

"Every subject!" he told me.

I didn't understand yet.

"I can help you with history."

"Okay, good."

"What period would you like to work on?"

"Start from the beginning. The school I go to, they teach me nothing."

I soon learned he had lived his whole fifteen years in a completely closed Orthodox Jewish community within the physical boundaries of Brooklyn but in an altogether different universe. He told me his classmates at his religious school barely spoke English. Their education was confined to the world of their belief. And yet, somehow, he had gotten the idea that everything he had been taught wasn't necessarily totally true. Somehow, innately, he had a sense of skepticism, an intellectual independence.

We sat down with the most remedial history book available.

We began in prehistory, before we were even human. I worried this would upset him, but he seemed to take the news of evolution rather well and asked cheerful questions.

A few minutes further into my abridged history of the

world, we got to the topic of early cave paintings. He asked me if this part took place six thousand years ago. Hesitantly, I told him it was more like thirty-five thousand or forty thousand years ago.

"In my community, they think the world is six thousand years old," he explained.

"I know," I said, and after some silence added, "Do you think the textbook is wrong?"

"No. I believe the textbook." He told me he thought maybe the root of the beliefs he had been brought up with was some kind of enormous historical misunderstanding.

As we continued I used the term *BC* several times without explanation. On the fifth or sixth time, he stopped me.

"What is *BC*? 'Before counting'?"

Awkwardly, I explained that I knew the year was 57-something-or-other in the Jewish calendar, but in the secular world, paradoxically, we use a calendar based on the Christian belief system.

He accepted this and we continued through the history of the world, defining whole millennia with single sentences, glossing over great civilizations, carelessly racing toward the present as afternoon turned to early evening. So much imperialism, so many revolutions. Almost all news to him.

Subtly I tried to signal my own, albeit much less intense, Judaism, but he saw me as Gentile until we arrived at the nineteenth- and twentieth-century immigration waves. I told him this was how my family came to America from Eastern Europe, to escape persecution.

"You are Jewish, too?"

"Yes," I said.

He was pleased. After a moment I thought I'd better add the catch "But I don't believe."

"I don't believe either, that's why I am here. My mother is very religious. I think my father is an atheist, but he will never say."

I had many questions but thought it improper to pry.

It occurred to me as we covered so many uprisings and revolts that he would have made a very good revolutionary. Here he was, learning the secular history of the world because he had dared to question the version he was being taught. He had decided the powers that be were wrong, and he took matters into his own hands. That is the story of so much of the history of the world, and the heart of so many national independence days.

Independence days are not biological like birth or death. They are not astronomical like the solstices and the equinoxes. But they spring from the same human impulses that fuel scientific discovery. Political revolutions and scientific breakthroughs require the same unwillingness to accept authority on blind faith. They are born out of the same question: *Why are things as they are?* They celebrate our ability to evolve in the figurative sense.

As with any abstract thing that we hold dear, we have to act on our belief, not just think about it.

For those of us who live in democracies and republics, there is a ritual we get to perform, one that many a martyr fought, bled, and sometimes died for: the casting of a vote. Around the world, on appointed days, a sticker or an ink-stained finger is evidence of one's faith in their system. If representative government were a religion, big national elections would be the iconic, marquee holidays, like Christmas. Local off-year ones would be the less-celebrated Maundy Thursday, remembered mostly by more devout, statistically older adherents. But as with all philosophies, our participation is actually required every day.

When I was young, growing up in the mainstream United States, I was also prone to question what I had been taught in school. Not necessarily because I had, like my student, some innate skepticism, but rather because that's how my parents raised me. I attended a good public school where I was not taught a particular religious doctrine. But despite the progressive politics of our town, some national chauvinism and blind patriotism made their way in. We said the Pledge of Allegiance, a daily mantra purporting a God and a place where there is "liberty and justice for all." The God part I'm dubious about but cannot be absolutely certain either way. The "liberty and justice for all" part is unequivocally a lie. And yet, millions of schoolchildren say this every weekday morning, from sea to shining sea.

It's hard to hold both the mythology of America, its aspirations and promises, and its crimes in my head at the same time. The war on nuance in our politics and culture tries to oversimplify complicated issues: *If you criticize the injustices in your country you must hate the troops!* or *How can you support that candidate when she's obviously not perfect?* Our fear of complexity, our inability to, as my dad put it, "tolerate ambiguity," is so often one of our biggest failings.

How do we deal with both? How do we face two opposites bound together in a single defining idea?

My parents were madly in love. I would walk into rooms and find them making out many years into their marriage. This was horrifying at the time, but in retrospect so lovely. They were very happy together and fought infrequently. One of the only things I ever saw my parents argue about was Thomas Jefferson. They both agreed that he was despicable for enslaving other human beings. And they both agreed that he had come up with some brilliant ideas about how to start a country. Their debates were not about if he was good or evil but about how to wrestle with his complex legacy. What do we do with nations and people who have both good and evil in them?

One year on the Fourth of July, when I was nine, my parents and I went to Jefferson's home, Monticello, now a historic landmark, where my dad conducted the annual naturalization ceremony.

Watching these fifty or so newly minted Americans from all over the world take the verbal oath of citizenship, I couldn't help being reminded of the spells cast in the stories I often read.

Aloud and under the right circumstances you say:

I hereby declare, on oath, that I absolutely and entirely renounce and abjure all allegiance and fidelity to any foreign prince, potentate, state, or sovereignty, of whom or which I have heretofore been a subject or citizen; that I will support and defend the Constitution and laws of the United States of America against all enemies, foreign and domestic; that I will bear true faith and allegiance to the same; that I will bear arms on behalf of the United States when required by the law; that I will perform noncombatant service in the Armed Forces of the United States when required by the law; that I will perform work of national importance under civilian direction when required by the law; and that I take this obligation freely, without any mental reservation or purpose of evasion; so help me God.

And then *poof*! You're an American! Sure, there are

other steps and tests to this ritual. You have to study, a kind of ritual unto itself. You have to learn the difference between the House and Senate and the names of the founding fathers, the history of our nation as we might wish it to be. But then, once you file all the paperwork and pass your citizenship test, you declare—like the founding fathers, like every sovereign government in history—what you now are, and lo and behold, you are transformed.

It's a very traditional kind of ritual. Just the kind of thing Van Gennep would have called a rite of passage. You start out separated from the group, in this case the United States of America. Then you enter the liminal phase, the naturalization ceremony, where you say the magic words. And on the other side you are united.

In his speech, given before the oath was taken, my dad told our new countrymen and countrywomen that the duty of an American is to question everything, especially authority, and to think independently. This is the very nature of democracy. As he and my mom had once written in an essay called "Real Patriots Ask Questions," he said that patriotism is not about blind obedience but about finding ways to make the system better. So he had some amendments for the oath:

And so it seems to me that part of the duty of citizenship is not to be intimidated into conformity, to be skeptical. I wish that the oath of citizenship that you are about to take in the next few minutes included something like, "I promise to question everything my leaders tell me." That would really be Jeffersonian. "I promise to use my critical faculties. I promise to develop my independence of thought. I promise to educate myself so I can make independent judgments." And if these statements are not part of the oath, you can nevertheless make such promises. And such promises, it seems to me, would be a gift that you can make your country.

If an idea cannot stand up to scrutiny, it should be discarded. That's the way of scientific discovery. And the pathway toward a more perfect union. There is a grain of this in national remembrances of uprisings, like the French holiday that marks the storming of the Bastille on July 14, as well as with days like Juneteenth, which commemorates the emancipation of those enslaved in the American South. The date, June 19, refers to the abolition of slavery in Texas specifically, but the holiday in its modern form honors all black Americans freed from bondage. Likewise, June is gay pride month because of an uprising over discriminatory police tactics at the Stonewall Inn in New York City's West Village. It paved the way for the gay rights movement and eventually for marriage equality. Acknowledging and enacting the need for progress doesn't mean one doesn't love their country.

The writer Sarah Vowell might have put it best. She writes about American history, often as influenced by her own travels across this great and terrible land. In one essay, she recounts how she and her sister, who are part Cherokee, retraced the Trail of Tears by car almost two centuries after their ancestors were brutally forced to leave their homes and make the trek on foot. As they drove, they listened to the greatest of America's rock and roll. "When I think about my relationship with America," she writes, "I feel like a battered wife: Yeah, he knocks me around a lot, but boy, he sure can dance."

Maruja very much identified as Inca. She spoke their language and often told me of the glory of her ancestor's empire. How did she reconcile that pride with the idea that the Inca were obliterated by the people who brought them the Catholicism she loved so dearly? I wish I'd asked her.

In the same way that religions can't help building upon the belief systems they replace, even the best-won revolutions cannot turn back time. They cannot reinstate the world as it was before the first ship arrived, before the first treaty was broken, the first village raided. They cannot undo the melding of cultures, ideas, belief systems, language, and genes. We can time travel, but not backward. Only forward one second at a time. One of my dad's most often-repeated quotes is "If you wish to make an apple pie from scratch, you must first create the universe." It applies to nations, religions, philosophies, and cultures, too.

I try to build my own philosophy, but it is, as this book attests, overwhelmingly the philosophy of my parents. If you have read their work, you know there are many familiar themes and concepts here. It's what I believe in, but it is not my original invention. Because of this, I found myself full of a kind of admiration for the Orthodox boy. He was doing something I never had occasion to do: rebel.

"I feel like I'm just copying you all the time," I said to my mother not too long ago.

"Or you could think of it another way," she said. "You are honoring thy mother and thy father."

Sure, there are ways in which I see the world differently from my parents, ways in which I disagree, times when I was an antagonistic teen, but I've experienced no rejection of their deepest-held beliefs. I've tested them in my head a million times, wondering, *What if I'm wrong? What if there is a God? What if I'll be somehow punished for not believing?* But I have never been able to conjure up a counterargument that made sense to me. In this way I fear I am not in a position to write about independence days and rebellions. On the other hand, adherence to our parents' views was not forced on us. They were always open to debate from their children. And many years of deep, long philosophical discussion have resulted in Sam and I both adhering to the basic principles they taught us.

Questioning something, exploring it, examining it, thinking of ways it might change for the better is a way of loving something. Barbecues and fireworks are all well and good for celebrating national independence. But I propose the real ritual should be teaching ourselves and our children to question our own preconceptions. Ask your children why things are the way they are. Ask yourself. Could things have gone another way? Imagine if the revolution had failed. Would we be taught what traitors Washington, Franklin, Jefferson, and Paine were in classrooms under waving Union Jacks? Or would some other revolutionaries have eventually picked up where they left off? Would slavery have ended sooner if the revolution failed? This kind of questioning, a version of the Socratic method—or maybe devil's advocacy—can and should be part of everyday life, but as with so much (giving thanks, buying your girlfriend flowers), we sometimes forget to do it if society does not appoint a particular day. Let our independence days become that appointed time to question, the moment when our minds break free from the absolute monarchies of "just the way things are."

Ithaca sits at the foot of one of the longest of the Finger Lakes. Together, from the sky, these look like some eleven-fingered deity scratched deep blue lines into upstate New York. Our beloved lake is named for the Cayuga tribe, members of a coalition of tribes called the Haudenosaunee, or the People of the Long House, who got stuck with the name given to them by French invaders: Iroquois. Not far from the homes Jon and I grew up in, there is a place that was called Coreorgonel, which in the Cayuga language means "where we keep the peace pipe." There is a plaque there that marks the slaughter of the entire village at the hands of the Revolutionary War general John Sullivan and his men, on the orders of George Washington. Among the Haudenosaunee, Washington was known as Town Destroyer. Jon and I have spent many a Fourth of July weekend swimming in Cayuga Lake, laughing, drinking, and not once stopping to think about what happened to the people who swam in it before us.

To the Zoroastrians, Alexander the Great was a devil who executed their priests. In Istanbul there are images honoring Atatürk everywhere, while in Armenia he is a mass murderer. Christopher Columbus has a national holiday in his honor, but for Native Americans he is the source of untold horrors and annihilation. When our daughter is old enough to grasp the basics of the history of the world, we will remind her often that history is and always has been written by the victors. As she questions history she'll have to question us, too. Test our ideas. Test what we've told her. We'll no doubt realize that there are ways in which we are mistaken. And we'll try to grow.

We all have our own agendas, doubly so when we record and retell history. Every omission, every emphasis is a decision designed to shape the reader's perspective, consciously or unconsciously. In my case, with my curious Orthodox pupil, it was conscious. I emphasized the parts of the textbook that talked about evolution, about the long history of our species, about how we were all originally from Africa. Those were the things that I thought were important. I wanted to bring him around to my side. I wanted to convince him that the walls between his community and the rest of Brooklyn, the rest of New York City, the rest of the United States, the rest of the planet, are artificial. We're all the same. That was the coded message I was trying to send.

I still think about the Orthodox boy a lot. I wish I'd had weeks to recount all of human history to him in detail. But I didn't know if he would ever come back for the next installment. Better to make sure we got in as many of the big battles, migrations, and revolutions as possible.

He did come back, but only once or twice more. In those subsequent meetings I learned he wanted to escape the fundamentalism of his community, but the organizations that aid in helping people out into the world don't accept minors. He had three more years until freedom, until his own independence day. Which has now come and gone. And I'm left wondering what big battles, migrations, and revolutions he has taken on since then.

*

chapter nine
_____

Anniversaries &
Birthdays

History doesn't repeat, but it rhymes.
—MARK TWAIN, *allegedly*

A few years ago, when my grandpa Harry was at the end of his life, I came home to Ithaca to be with him. He was 99, but until very recently he had been preternaturally, almost jarringly youthful. So much so that despite his very advanced age our whole family felt confident he still had a good number of years ahead of him. His mother lived to be almost 102, and that was thirty years ago. He could surely make it to 103 or 104. He'd only stopped driving a few months before. But the reality was that his kidneys were failing and he decided against dialysis. He was a lifelong lover of things of very high quality and in the end he chose quality over quantity. I spent his last few weeks in Ithaca relishing time with him as he slipped away. It was very difficult knowing it was a one-way street leading, sooner or later, to loss. I was an emotional wreck but tried to be positive for my granddad. Jon came on the weekends, which helped. Grandpa Harry was increasingly mixed-up and sleepy, in and out of lucidity. One day he turned to me and asked very frankly, "Who's having a baby?"

"I don't know," I said, smiling.

Jon and I had been trying to get pregnant for about six months. I thought maybe I would take a test in a few days and get to tell my granddad that I was pregnant before he left us. Two days later he slipped into a state somewhere in between life and death. I got my period that morning. Then he died. I was devastated and heartbroken. I had thought losing my grandpa would be easier if I was able to tell him he had a great-grandchild on the way. But that was not to be.

And instead of planning for a baby, we had to plan a funeral on October 28.

English is missing a concise, simple word that means "death day," or the anniversary of someone's death. But Yiddish has one: *yahrzeit*. In Judaism, there is a traditional candle lighting on the *yahrzeit* of a loved one, using special candles that burn for twenty-four hours. It's the paradoxical antithesis of a birthday candle that gets blown out right away. From the time I was small, my mother taught me to light *yahrzeit* candles to commemorate Rachel, Tillie, Benjamin, and other ancestors I had only known from pictures and stories. As I grew and lost people I knew and loved, I had my own grief and lit my own *yahrzeit* candles. I love this tradition because it feels like a miniature version of dead stars that appear to twinkle even after they're long gone. When someone we love dies here on Earth, the news can take a long time to absorb. It's shocking. Something about that small light gives us the appearance that they're not all the way gone yet.

We do something like this on a countrywide scale, too. At Arlington National Cemetery, the gravesite of President Kennedy features an "eternal flame." This is a misnomer, since nothing is truly eternal. If the sun will eventually go out it's safe to say Kennedy's fire will, too. But it was a way to comfort a traumatized nation after his assassination, a way of reassuring everyone that some part of him continues on.

Even birthday candles appear to have ancient and astronomical history. The Greeks had a moon goddess, a huntress, called Artemis. She was, as the moon still is, associated with wild animals and female reproduction. Euripides called her the "easer" of birth. Members of her cult in Athens may have been the first to put "little torches" atop a flat cake. They did this not to celebrate birthdays but to reflect Artemis's role as "light bringer" and mimic the glow of the moon. Perhaps, because she was the goddess of birth, this strange, obscure custom somehow slid down the generations to become a *birth*day ritual.

But why light a candle on the same date at all? I miss my dad, my grandparents, Maruja, and the other people I've lost all year round. Likewise I am getting older all day, every day, not suddenly aging a full year on my birthday. What is it about doing things annually that feels so right?

Somehow we humans got the idea that the best moment to really meditate on an event—a birth, a marriage, a death, a battle, a coronation, an inauguration, anything good, terrible, romantic, auspicious, historic, or otherwise memorable—is when the Earth is back in the same position it was when the thing happened. This is astronomy at work. On anniversaries and birthdays, we are in the same place in relationship to the sun. However, the whole solar system is actually moving through the galaxy every 225 million years. So you're kind of in the same place, but everything is also totally different.

I suppose that's a pretty good analogy for what birthdays and anniversaries feel like. Imagine your childhood birthday parties, for example. There are reliable tropes year in and year out: the cake, the games, maybe the paper hats. But you and your friends are changing.

In the ancient world, the birthdays of powerful people were marked, in some cases ordinary people too, but it was between the Victorian era and the 1950s that the concept of an annual party, in particular for children, trickled down from the aristocracy to the rest of us.

For much of history most people didn't know what day they were born or what day it currently was, which made celebrating this kind of thing pretty tough. But once we started keeping track of when things happened we became obsessed. Every day of the year there are countless historical and personal events being commemorated around the world. In the United States we make holidays out of birthdays like Martin Luther King Jr. Day and Presidents' Day. The birthdays of kings, queens, and dictators are celebrated around the world. There are religious holidays to commemorate the births of founders or reformers, like that of Joseph Smith for Mormons or Mahavira for Jains, not to mention Christmas. Although they are rarely celebrated on the icon's actual birthday, they are a way of extending our connection to those we admire after their deaths.

It's also a way to feel that we know our coordinates in the universe. We want to feel acknowledged, marked, by the vastness. It's this desire that makes astrology so popular.

Something happened the month after Grandpa Harry died. Or rather, something didn't happen. My period didn't come. I waited a week. Tried not to get my hopes up. Maybe it was just late. But it wasn't. I was pregnant. I still feel a rush of emotions when I remember looking at the pregnancy test, Jon and I slack-jawed in amazement.

Using the date of my grandfather's *yahrzeit*, the first day of my last period, we calculated when our baby would be due: August 4. We couldn't believe it. Not only was it the birthday of someone Jon and I deeply admire, but even more amazingly it was twelve years *to the day* since the hot New York night when our friendship finally turned romantic.

This seemed miraculous. I couldn't shake the temptation to ask myself, *What does this mean?* What were the chances? It seemed important. It was hard not to get very attached to my due date. I know babies rarely arrive on schedule, but I really wanted our baby to be born on August 4. Not a day early, not a day late. That day.

However, I had some complications, and my obstetrician decided I should be induced a week early, depending on when they had availability at the hospital. The nurse gave me a piece of paper that said we should call in on July 30 and, provided that they weren't too busy, come in and have a baby.

A July birthday seemed so different than what we had imagined. I felt devastated. I still to this day don't know exactly why. Maybe it was just the change in plans, the feeling I wasn't in control. Or maybe it was some deep unscientific belief that the day she was born would determine who she was.

Here's an awkward exchange I've had countless times throughout my life:

PERSON: "What sign are you?"
ME: "Scorpio, but I don't believe in astrology."
PERSON: [rolls eyes] "Oh, that is such a Scorpio thing to say."

Sometimes the person gives lots of anecdotal information in order to convince me: "But I'm just a classic Taurus, so it's true for me." "All my best friends are Geminis." "Every terrible boyfriend I've ever had was a Libra." "My horoscope is always right." Sometimes I explain that these conclusions are unscientific, the sample size is too small, they're counting the hits and not the misses. Sometimes I keep my mouth shut. Either way, we usually both remain unconvinced by the other and move on.

But once my belly began to give away the fact that I was pregnant, every single day someone would ask me when I was due. I was astonished at how often the conversation went like this:

PERSON: "When are you due?"
ME: "August fourth!"
PERSON: "Oh, a little Leo!"

This exchange was more disconcerting for me than the one where I am accused of being a classic Scorpio. Make whatever arbitrary judgments you'd like based on my birthday; I'm already me. But the little being in my belly? Her identity was not yet formed, but strangers felt totally confident that they knew what she would be like. Truth be told, I have no idea what traits are supposedly associated with being a Leo. They could be great. Maybe I should have been hoping to give birth to a "quintessential Leo," but these presumptions about my baby's personality bothered me.

They bothered my mom, too. "It's like any other prejudice, a *prejudgment*. You know one thing about a person and you think you know who they are," she told me. She's right. It's a kind of stereotyping.

To the best of my knowledge, no star sign has ever been subjected to the kind of discrimination that has all too often come with skin color, gender, ethnicity, sexual identity, or religion, but there is a parallel. Saying "I know what kind of person you are because I know one thing about you" smacks of the same laziness of assumption that fuels the isms.

Astronomy and astrology used to be one thing: the real observations of the stars and planets intertwined with our fears and wishes about what they meant. They began to be pulled apart in the seventeenth century. Ideas were tested. Some could stand up to scrutiny, and some could not. But the same instinct still pulls us toward both astronomy and astrology.

My parents took great pains to teach me, and many millions of others, that we *are* truly and literally connected to the universe. We don't need to make up reasons why or how, or look to the specific movements of the planets at the moment of our birth to determine who we are.

Despite my total lack of faith in astrology, I did care when my baby was going to be born. If she came on August 4, I thought, she would somehow fit into a pattern. Finding patterns is something human brains are very good at because it's a huge evolutionary advantage. For example, newborns who recognize the pattern of a human face—namely two eyes, one nose, one mouth, in the particular arrangement to which we are accustomed—smile when they see one. This reaction bonds them with the owner of the face they are looking at, and said face owner is then more likely to take better care of the helpless baby. Babies who didn't make such strong bonds with their grown-ups were not as well taken care of and survived less often. Because of this, face-recognizing genes endured and multiplied. So much so that now we see faces where there are none, like the man in the moon. We're so good at seeing patterns, we almost can't stop ourselves.

This is true for meaning, too. Sometimes something so astonishing happens that we can't help but believe there is a message in it. The desire to read into these moments is almost overpowering, intoxicating. My parents described this as being a "significance junkie." I am as guilty of this as anyone. My March Fo(u)rth obsession is evidence. And like everyone, I have had experiences that haunt me with their improbability. For example, once I was walking down a street in the West Village with a girlfriend. She was telling me about a man she had been in love with years earlier. She had been an assistant on a movie and he was a famous actor. The romance had never come to fruition, but the feelings lingered. She had seen him weeks before, after years apart, in the park with his pregnant love. When my friend told me, I said, "Isn't it amazing how as soon as you stop thinking about a man, he appears?" and then, at that moment, that very man and his glowing girlfriend materialized in front of us. We couldn't explain to them why this was so shocking and instead blathered hellos and goodbyes before they went on their merry way.

And then two men carried a skinned goat into a nearby bodega.

We were flabbergasted. It was so strange, so surprising, that we could not shake the feeling that it was some kind of omen.

Still now, years later, I get chills when I think about it. Of all the many coincidences of my life, this one stands out. It felt like being inside a novel, albeit one written by a very heavy-handed author.

Considering all the events and moments that make up a life how could there not be at least some that felt like serendipity? Even if all coincidences are completely unplanned, simply the by-products of statistics—like the proverbial monkey that types Hamlet—that doesn't have to make them any less exciting. It's beautiful that our brains have evolved to recognize patterns. It's one of our greatest strengths. It's what allows me to communicate ideas to you with these squiggly little symbols you're looking at right now. It's what allows us to understand concepts like mathematics and physics, it's how we manage to engineer buildings that don't collapse and planes that stay in the air. So in a way, there is a "grand novelist." It's us. We are the ones writing this epic saga, pulling out the plot points from the scenery.

At a dinner one night, an acquaintance told me a dead friend visits her and her husband in their dreams on the same nights. There's no question that having a similar dream as your husband is beautiful, no matter the explanation, I say.

"You really believe that's just random?" someone else asks, frowning.

It's the word *believe* that really stands between the two positions here. I don't believe these events are orchestrated by some larger force, because I've never seen any evidence to support this theory. I'm not swearing on my life that it couldn't possibly mean something. It is possible. I just withhold judgment—and belief—until there is proof.

Or at least I try to. That thought crosses my mind, too: *This is too amazing, too surprising. What are the chances?* I get that spine-tingling thrill. I also see patterns where there aren't any. Let's not forget, I'm the one who claimed men always appear as soon as you stop thinking about them. Talk about counting the hits and not the misses.

I strive to just enjoy that magical moment we all experience when, say, you're thinking of an old friend and they suddenly call. The urge to see a pattern is strong, even biological, but I also find pleasure in the idea of a world of total accidents. If these coincidences are orchestrated by some supernatural force larger than the human brain, well, then they were inevitable. But if it's truly random, if it's the one-in-a-million shot that lands, that is, to me, more special.

I had put off reading the bulk of my parents' books for years. I wanted to have some part of my dad to still look forward to, some part of him in my future. When I was pregnant, a whole bunch of their works were released on audiobook. I love being read to, I think because my parents read to me so generously when I was a child. And something about being on the brink of motherhood, about to meet a new person who would be one-quarter my dad, made me feel ready to dive into some of the books I had not yet read. Most of the audiobooks were performed by actors and members of my family, including my older half-brother

Nick and my sister-in-law Clinette, but there were chapters here and there that had been recorded, decades earlier, by my dad.

As I went about my pregnant business in my third trimester, I had my dad's warm, familiar voice in my ear thanks again to modern technology. On a hot early summer day I walked through Boston's Chinatown toward my local drugstore and heard my father say:

My grandfather was a beast of burden. I don't think that in all his young manhood Leib had ventured more than a hundred kilometers from his little hometown of Sassow. But then, in 1904, he suddenly ran away to the New World . . . We know nothing of his crossing, but have found the ship's manifest for the journey undertaken later by his wife, Chaiya—joining Leib after he had saved enough to bring her over. She traveled in the cheapest class on the Batavia, *a vessel of Hamburg registry. There's something heartbreakingly terse about the document: Can she read or write? No. Can she speak English? No. How much money does she have? I can imagine her vulnerability and her shame as she replies, "One dollar." She disembarked in New York, was reunited with Leib, lived just long enough to give birth to my mother and her sister, and then died from "complications" of childbirth. In those few years in America, her name had sometimes been Anglicized to Clara. A*

quarter century later, my mother named her own firstborn, a
son, after the mother she never knew.

It was part of the introduction to his book *Pale Blue Dot*, a meditation on our species as wanderers, migrating all across the planet and maybe someday beyond it. Its message is applicable to every human who has ever lived, but it felt like a secret meant just for me. What passersby must have thought when they saw me, doing errands, extremely pregnant, wearing headphones, hysterically crying.

Traditionally, Ashkenazi Jews do not name children after the living. And even when naming after the dead, the practice among Reform American Jews is more often than not to just take the first letter of your dearly departed's name and go from there. I suspect this is to ease assimilation. My father had been named Carl in this tradition. To honor his mother's mother, Chaiya, his parents could have named him Chaim, the male form of this name that means "life." But they didn't. They wanted him to be American, not Russian, not Yiddish. I knew this growing up, and from the time my father died I imagined giving a child a name that honored him with the letter *C*. When my grandpa Harry died twenty years later, Jon and I decided we must find a way to honor him with a letter-*H* name, too.

A few months later, while I lay shaking from anesthesia and shock on the operating table of the maternity ward of one of the greatest hospitals in the world, surgical lights in my eyes, I thought about my great-grandmother Chaiya. I had not planned to have a C-section. In fact, I'd been trying to have my daughter the old-fashioned way for two days, but her heart rate wasn't coming back up after each contraction. Together with the doctors, Jon and I decided I needed surgery. I'll never forget Jon looking down at me, tears in his eyes, just before I was wheeled into the operating room. "We get to meet her soon," he said.

Not long later, I got my first glimpse of her. From my vantage point on the operating table, I couldn't quite get a good look at her gorgeous little face. But I immediately spotted something familiar. She had my same oddly shaped big toes, which I got from my dad and maybe once belonged to some long-forgotten ancient forebear. Once they put her in my arms I could see that she was perfect, with a full head of hair and a strong heart.

It was August 1.

I lost a lot of blood. I was dizzy, very thirsty, stunned, and profoundly happy. The whole experience was terrifying, but I knew I would survive.

We named our girl Helena Chaya. Helena means light, from Helios, the Greek sun god, my old friend from the cover of my book of mythology. For this child who entered the world in the summer, who lit up our universe. We decided Chaiya was a better middle name than a first and we spelled it the more common way, without the *i*. Like my father's parents, we decided to give our child a first name that would make life in the society she was living in a little easier, something that allowed her to avoid constantly correcting people's pronunciation (the *ch* makes that fricative, Yiddish *h* sound).

In the days and weeks after Helena was born, as I watched the long scar across my abdomen heal, I thought about how many babies over the course of history would have never made it under those circumstances. How many women bled to death giving life, or died from infections days later, like my great-grandmother and Maruja's mother. I thought about how many women still die today because they do not have access to a world-class hospital.

There is a direct correlation between science and survival. Medicine is a branch of science. Access to it keeps us alive in situations when we would have otherwise died. Infant mortality is still heartbreakingly high where war or poverty makes medical care scarce. But even the most powerful medieval queen could not have saved a baby in desperate need of antibiotics, an incubator, or emergency surgery in a clean, well-lit operating room.

Whatever else may have played a part in our survival, there is no question that Helena and I are the beneficiaries of the scientific method.

It's this rejoicing at survival that's at the heart of every birthday party. For most of history, survival was harder. Children perished. A first birthday was not a given. Each passing year, each step closer to adulthood, was a relief not to be taken for granted, something truly worthy of celebration. That's what a birthday is: the realization that time is passing but we are still alive.

Months after Helena was born, when new characteristics, elements of her personality, and the contents of her genome seemed to be revealing themselves daily, I bought a book about heredity. The title attracted me: *She Has Her Mother's Laugh*. It reminded me of my connection to Rachel. As a kid I had been obsessed with chimeras, mythical creatures that were mixes of two different animals, like the Pegasus or the Sphinx. In Carl Zimmer's book I learned that, in a way, I might be just such a creature. In 1996 the first scientific paper was published that showed that women not only pass on their genes to their babies but that their babies leave cells behind that become part of the mother. It's called microchimerism. Helena may have left some part of her genetic self inside me in a literal, provable sense. She is emotionally part of me, but she may be physically part of me too, as Jon and I and all our predecessors are part of her.

As Helena grows we'll celebrate her birthdays with cake and parties, or whatever she pleases, but every year, while she opens her gifts, some part of me will be back in that operating room, under the bright lights, watching her take her first breaths, entering the world with all the qualities be-queathed to her by all her ancestors. Every one of whom was crucial to getting to that moment.

When I was pregnant my mother told me that when I saw my child for the first time I would think, *Of course! It's you!* She was right, as she so often is. All my life I've heard people say, "I feel like I knew so-and-so in a past life," casually in-voking reincarnation and mystifying me. I've certainly had the experience of giddily clicking with a new person in a chorus of "Me too! Me too!" but this whole idea of a past life seemed incredibly dramatic.

Until I saw Helena for the first time. She was so tiny and red and cute on that first day of her life, her literal *birth*day. She was so deeply familiar to me that I couldn't believe I'd only just met her. I felt like I had known her all my life. And in a way I had. In her I could see the echoes of Jon and my-self, of my parents and of my in-laws and their parents. Go-ing back beyond the horizon of people whose names I'll never know, to the earliest humans and before. What tiny charac-teristics might Helena have inherited from some hunter or gatherer? Does she make their faces? Their sounds? Have their talents? Their idiosyncrasies? Or is all that too diluted by time to be passed on whole? I don't know. But in the days and weeks after her *birth*day, Jon, my mom, my friends, and I all marveled at the uncanny familiarity of a particular smirk—pursed lips, nostrils flared. It was unmistakably

Grandpa Harry. Soon she started making his signature sigh when she wanted attention. It's this sort of experience that no doubt lends credence to the belief that babies are actually the reincarnation of their ancestors. It was no less astonishing to me, no less beautiful and no less reassuring, just because there was a biological explanation. If anything, that made it more stunning.

chapter ten

Weddings

*Call it what you will—Happiness! Heart! Love! God! I have
no name to give it! Feeling is everything, name is but sound
and smoke...*

　　　　—JOHANN WOLFGANG VON GOETHE, Faust

*Raise high the roofbeams, carpenters! [God of marriage], sing
the wedding song!*　　　　　　　　—SAPPHO

I was still recovering in the hospital on August 4, the
twelfth anniversary of Jon's and my first night together.
It's one of many anniversaries we mark. But our wedding
anniversary is the one we celebrate in the biggest way. Helena was about six weeks old when our fourth rolled around.
I thought about it so differently now that we had a child together, a child who might someday see that day as integral
to her very existence. When I was small I spent hours por-
ing over my parents' wedding album. I couldn't believe I

had missed something so important. I wanted to time travel there as well.

My parents happy, romantic marriage reflected positively on the institution as a whole. They brought me up to be skeptical, certainly, but not cynical—least of all about love. "Give them the benefit of the doubt," my mother would tell me and others when they were fighting with their partners. She taught me to assume the best.

Because of my parents I always knew I wanted to get married. But after my father died my feelings changed. I still wanted to *be* married, but the *getting* married part seemed increasingly like it was going to be more painful than joyful.

Around the time we became serious, Jon invited me to be his date to a friend's wedding. At the rehearsal dinner the bride's father went into great detail about how much he loved his daughter and how precious their relationship was. It was beautiful, but I was jealous and heartsick. I excused myself to the ladies' room to weep. I was sure a wedding without my dad would be too awful to ever face.

I imagined the rituals leading up to a wedding would mostly be shared with my mother. I knew she would help me plan the wedding and take me dress shopping. I have several friends who lost their mothers in the years and decades before their weddings, and for them, these were the rituals that felt painful. For me it was the idea of not having my father on the wedding day itself. The idea of him "giving me away" is ridiculous and antiquated. I am my own grown-up person. But not having him there to walk me down the aisle, despite the original misogyny of the practice, felt crushing. Why would I ever invite everyone I know to come stare at the horrible empty hole in my life?

After that rehearsal dinner Jon and I sat on a bench at the foot of Central Park and for the first time I let all my grief out in front of him. He didn't tell me a happy lie, that everything would be okay. He didn't tell me to look at the situation differently. He didn't tell me not to have my feelings. He just held me and listened, tears streaming down his own face. For that—and for a million other reasons—there was no question about my answer when, years later, he got down on one knee.

But after our joyful tears and champagne, there was a celebration to plan.

Months later, Jon and I were in the lobby of a midtown hotel with our wedding planners and the lead singer of our wedding band.

"Are you doing a father-daughter dance?" the singer asked.

I frowned.

"No, my dad passed away," I told her.

The singer flinched. I wanted to alleviate her embarrassment, but the moment passed before I could lie and say it's okay.

She smiled warmly. "Your father will be up there watching over you on your special day."

This kind of sentiment makes things more awkward for me. I don't want to pretend I believe in an afterlife, but I don't want to be rude or start a philosophical debate. Jon squeezed my hand. I made a vague, noncommittal sound and tried to steer the conversation toward something else, anything else.

Jon and I met in middle school in the mid-1990s, but it took a long time for us to realize we liked each other. A really long time. For many years we were just acquaintances. We went to the same parties but can both only remember speaking to each other a handful of times. Most memorably when I parked my mom's car on his foot. I had just (barely) gotten my license and was too distracted by the fact that a cool senior was talking to me to realize I was slowly, almost imperceptibly, rolling forward. Imperceptibly to me, very perceptibly to Jon, the cool senior. He was so calm and nice about it I should have fallen in love with him right then in his father's driveway. But it took a while longer.

After we both graduated from college he moved down to New York, where I had been at NYU. In those days, before Google was the answer to everything, it was nice to know someone in a new city. Soon we became good friends. And I developed a terrible, debilitating crush on him. Eventually he came around. Then we pretended we were just "friends with benefits" for almost a year, while everyone we knew could plainly see that we were falling in love.

My father couldn't walk me down the aisle, no matter how much I wanted him to. I had to find some other way to incorporate him in our wedding. I decided we needed to get married somewhere my dad and I had been together. Not because I believed his spirit would still be there, but because my memories of him would be.

It turns out there aren't that many places one goes as a kid with their dad that also happen to be wedding venues. But on a trip home to Ithaca, early in our engagement, we drove by the Herbert F. Johnson Museum of Art, a modernist structure designed by I. M. Pei on Cornell's campus. When I was small my father would take me there to see the Giacomettis, the Calders, and the Japanese scrolls. In 1996 it was the site of his memorial service. In 2002 my brother Nick and his wife Clinnette had their wedding reception there, too.

But when we went to check it out, Jon and I saw something brand-new. My mom had mentioned it in passing, but seeing it with our own eyes floored us. On the ceiling that jutted out several stories above the large balcony, there were lights. Twelve thousand lights. Constantly changing, enormous and awe-inspiring, creating patterns that evoked the depths of space. The artist Leo Villareal had created it and called it *Cosmos* as an homage to my dad's work and the grandeur of what he called "all that is, ever was, or will ever be."

It was breathtaking.

What unbelievable good fortune for me, I know. How lucky and unlikely that there should be an enormous art installation that could serve as a perfect metaphor for my deepest emotional needs. Doesn't really apply in most situations, does it? But in a way it does. All our best rituals are a kind of performance about what we need or want most. Sometimes they are so on the nose they barely qualify as art—for example, a kiss at the end of a ceremony to signify the sealing of the union. Sometimes the traditions are so open to interpretation that they mean something different to each of us. And sometimes the origins of the tradition are so old and so popular that we abide by them despite not understanding what they really mean. Throwing rice at the newlyweds after the ceremony was once a fertility rite, a way of trying to harness some good luck in conceiving many babies, but who thinks of that now, standing outside a church, sprinkling your friends?

Besides rice, other foods have come to be perceived as

good omens for baby-making. The egg is a popular one. Muslim brides and grooms have been receiving them for centuries. In Hungary it's wheat, which gets braided into a bride's hairpiece. It doesn't have to be food. Greek wedding parties have a tradition of laying a baby in the marriage bed before the bride and groom get there.

And like marriage itself, it's not all about getting pregnant. There is a German tradition where the bride and groom use a two-person saw to split a large log in half, a kind of marital "team-building exercise" to prepare them for a lifetime of cooperation.

Symbolism around marriage is so tightly woven into our lives we don't even realize it's there. You both put a metal circle on your fourth finger because of an ancient erroneous belief that there was a special vein in only that finger that led to the heart. And then, when people out in the world see you have this metal circle, they know not to try to have sex with you. In the abstract it sounds bizarre, but it's a small, elegant way of saying, *I have found my true love.*

There are many other examples. A white dress for purity, a fad started by Queen Victoria. Something old, something new, something borrowed, something blue. Tossing a bouquet or jumping over a broom. Planning even a secular wedding highlights how much we humans delight in inventing and keeping traditions. Some felt like impositions from bygone eras, like wearing a veil. I rejected this one because I couldn't separate it in my mind from the idea of an arranged marriage to a stranger waiting at the altar to see my face for the first time.

In all my attempts to be modern, my wedding did bring out some deep, hidden, traditional part of me. Jon and our secular rabbi had to convince me to do pictures and the "first look" before the ceremony.

RABBI: "You don't want to miss cocktail hour!"
JON: "The people in the back of the room will see you before me!"

But I wouldn't budge about where we slept the night before the wedding. For some reason, inexplicable to me, after six years of living together I was determined not to spend that night in the same bed. I, along with my best girlfriends, slept at my mom's house.

At five thirty on the morning of my wedding I woke up tingling with excitement and somehow forced myself back to sleep for an hour. At six thirty, I got up, snuck through the house, swiped my mother's car keys, and left.

When I returned, my mother and all my girlfriends were up. No one was worried that I was a runaway bride. My girlfriends had assumed I'd decided that not spending the night before the wedding together was a huge mistake and that I had gone to Jon's hotel room for one last premarital romp. Not at all implausible, but that's not where I was. Only my mom guessed correctly. I had been at the cemetery visiting my dad. It's a beautiful cemetery and it was a gorgeous, sunny late September morning. I sat in the grass in front of his headstone and wept. And talked to him. I do this whenever I go there. Not because I think he, or my nearby grandparents, can hear me. I don't. It's not for them. It's for me. It's a reminder that even though they're not here anymore, they were once and I still love them.

It was catharsis—the Greeks again. I needed to cry it out. I needed to feel this pain so I could experience the pleasure of the rest of the day. Crying on your wedding day is very common, and for a wide variety of reasons. Sometimes it's even ritualized, like for the brides of Tujia, who live in central China, where they are obliged by tradition to force themselves to cry in front of family. Like an acting student, the bride practices for weeks, sometimes years. As the big day approaches, her female relatives join in. Backed by an ancient legend, the idea is to highlight the joy of getting married by contrasting it with an opposite emotion. This is similar to what I was doing.

I don't know if any civilization on Earth has ever believed that on the morning of one's wedding day, before their attendants awaken, one must go to the burial place of their ancestors and wail. Cry out to them, feel all the pain of loss, alone. And then return ready to be anointed with oils and perfumed, dressed in their customary gown, the traditional color of virginity (cohabitation notwithstanding), and be made up to marry. For me, it was as good a ritual as any I've ever read about. I got all my tears out and felt better, able to focus on all I did have to enjoy rather than on what I had lost.

We had our ceremony at the top of the museum, overlooking Cayuga Lake and our whole hometown. My mother and Grandpa Harry walked me down the aisle. Despite our nonbelief, Jon and I incorporated three monotheistic traditions into our ceremony. We stood under the chuppah, the Jewish wedding canopy. Like so many traditions the meaning is open to interpretation, but we thought of it as a metaphor for the marital home: the roof shelters and unites the bride and groom, but it cannot isolate them. The four open walls welcome their families and friends, the rest of their community, and all the ideas and beauty that the great wide world has to offer. We lit a unity candle, a twentieth-century Christian tradition, designed to be a metaphor for the coming together of two lights into one shared flame. At the end of the ceremony, Jon stepped on the traditional glass (in this case a lightbulb) wrapped in a piece of cloth, to shouts of "Mazel tov!" There are dozens of theories about what this act symbolizes, but the most romantic is that it will be as impossible to tear the bride and groom apart as it is to put the glass back together.

We also incorporated something we loved from a poly-theistic tradition, a Greeks myth that moved us. The story comes from Plato's *Symposium*, a work of fiction. It's attrib-uted to Aristophanes, who was both a character in *Sympo-sium* and a real person, but there's no evidence he ever really told this story. Nonetheless, it goes like this:

Originally, there were three kinds of human beings—male and female, and a combination of the two called the "androgynous." People back then were completely round, with four hands, four legs, and two faces on a circular neck. They were ambitious, and tried to usurp the gods. Zeus planned to retaliate by cutting each of the humans in two, creating a species with two legs, two arms, and one face each. But the new half creatures were miserable. They missed their other halves. Some men were split from women, some men from other men, some women from other women. Whoever they were split from is who they sought. The half creatures would embrace each other, trying to reconnect, hoping to grow back together. They were so busy doing this they died from hunger and purposelessness. Zeus started to feel guilty. He reconfigured their bodies so that when they embraced they could have the satisfaction of reuniting into one creature, through sex. Love is the call back to our origi-nal form. Love heals the wound of human nature.

That was written 2,400 years ago. I don't know how many Greeks literally believed this story or, for example, that the Gods physically lived atop Mount Olympus. But you don't have to believe in Zeus to find beauty in their lore. It felt true to us—not literally true but as a beautiful metaphor that required no dogma and no faith. As with, say, a great sci-fi movie, no one believes they're watching a true story. *E.T. the Extraterrestrial, 2001: A Space Odyssey, Star Trek,* and the Star Wars franchises don't reflect a unified universe, but each still provides some insight about reality. So we included Aristophanes's myth, no differently than a poem.

There was one other small gesture we invented for the occasion. Not monotheistic, not polytheistic, not atheistic, just particular to us. Since my father couldn't walk me down the aisle, we wrapped an old necktie of his around my bouquet. Likewise, on the table where the unity candle was displayed, we placed a small lace headscarf Maruja had worn to church as a young woman. She gave it to me with the instruction to wear it to *sinagoga* (synagogue). I almost never go, so it lives in my bedside table. But at the wedding I could incorporate it into the kind of setting she had hoped for. These pieces of fabric did not bring them back. They did not convince me they were there in some supernatural way. But they brought to the foreground more of my memories of them, which still live on in the grayish folds of my brain.

After cocktails and speeches, the band played into the autumn night air and Jon and I danced under Villareal's ever-changing sky. There we took part in my favorite custom of all, the hora. I could argue it's the single best element of Jewish culture. I'll never understand how it hasn't been co-opted by everyone on Earth. It's the perfect metaphor for a wedding. Your loved ones encircle you, dancing, swirling around you, lifting you onto chairs and tossing you skyward. Just before I was about to be boosted up, one of my lifelong friends whispered in my ear, "Don't worry, you'll be fine." It was exactly what I needed to hear before being launched into the wild blue yonder. At college I had studied Dionysus, the ancient Greek god of wine, fertility, and frenzied, wild parties, whose cult honored him by achieving ecstasy through dance. I had thought of his worshippers on more than a few nights out in my twenties, in the crowded, loud, hot nightclubs of New York and London. But I never reached that divine state until I was up in that chair.

A wedding is not a biological or astronomical event. It's social, something we created for ourselves because we decided it was important. Marriage itself has had so many permutations and incarnations that any definition leaves something out. Some marriages last a lifetime, others barely through the reception. Some are open, some symbolic. According to marriage historian Stephanie Coontz, the marital arrangement "found in more places and at more times than any other" has been the union of one man and more than one woman. It's a very good system for producing as many children as possible, which historically has been a main goal of marriage. However, sometimes we do it the other way around, as in parts of rural Tibet where women commonly marry sets of brothers. There have always been long-term relationships between same-sex couples, and in ancient Rome and among tribes from North America to West Africa, there are instances where the society has recognized them as valid and legitimate marriages.

Today, in Shia Islam, there are temporary marriages (some even as short as an hour), which allow for automatic annulments afterward or renewal if it's going well. In theory, women must wait three months between marriages (which applies no matter what kind of marriage they're in or how it ends), but not everyone does. These seem to be quietly increasing in popularity in the United States and Iran. They're controversial. Are they like a free trial period? A preplanned divorce? Or a loophole to allow for prostitution? Or a perfect way to balance traditional theology with modern life? These are questions for Quranic scholars, but it is telling that the word for this practice, *nikah mut'ah*, is sometimes translated as "pleasure marriage."

How people marry is also vague and changing.

In medieval Christendom, marriages between the nobility were well-orchestrated mergers. For commoners, however, some combination of cohabitation, consummation, or just the utterance of a promise to each other was often all it took. The lack of any official record keeping or a single authority on these vague, informal commitments made for some complicated "he said, she said"s. So the Church set some ground rules. These included telling people beforehand (so they could object if need be) and having a ceremony in a church. It's then, at the Fourth Lateran Council in 1215, that the ritual of getting married starts to take shape in the West. You couldn't just say you were married anymore. You had to say it a certain way, in a certain place, to make it so.

The ritual has evolved, of course. We can use it to express something different from "This woman now belongs to this man." We have adapted it to mean something more like "Here are two equals who choose to be together, to make each other happier, better, less alone."

No matter when or where it happens, getting married feels like a huge leap into the unknown. But as long as you're marrying the right person, and you have the love and support of your family and friends, you'll be fine. Up in that chair; looking at the beautiful and good man I married and at the legacy of the beautiful and good man I lost, I felt not that everything was perfect but that everything was a poem, precise and sublime in its depth and truth.

Now, in our bedroom, we have a painting of the moment

that Jon and I were lifted up under the *Cosmos*, ecstatic with joy. The painting was a wedding gift from one of my oldest, best friends, painted by another close friend. I wonder now if Helena will someday look at it and feel that she should have been there, the way I felt looking at my parents' wedding album. There is so much that happens before and after each of us is born, we must relish the things that happen while we're here. We knew that we were missing my dad that day, and others once with us and now gone, but we didn't realize we were missing someone yet to come.

*

chapter eleven

Sex

Pleasure clearly is not evil but good . . .
—BARUCH SPINOZA

Folks, I'm telling you,
birthing is hard
and dying is mean—
so get yourself
a little loving
in between.
—LANGSTON HUGHES

C an you believe sex makes people? What! As long as I
live I don't think I'll ever stop being astounded by
this. Like so much in nature, when we take a step back and
describe it to ourselves it sounds like magic. I don't recall a
specific conversation when the proverbial birds and bees
were explained to me, but I do know I had a children's book
that was meant to shed some straightforward, scientific

light on the subject. The illustrated sperm had top hats, I remember, and an orgasm was described as something like the feeling of sneezing when you really, really have to sneeze but "much better" (which I learned a long time later was a comparison previously made by Alfred Kinsey). I remember being fascinated by this book but not what it felt like to learn this information. It's so hard to remember what it's like to not know something that eventually becomes so central.

The ritual of sex fulfills so many of the promises of religion: the miracle of creation, a feeling of transcendence, a chance at a kind of afterlife as children carry on your DNA. These are the qualities we usually ascribe to the divine, not the profoundly human.

What if instead of sex being associated with sin, with dirtiness, with secret shame, it was considered part of the glory of nature? If we set out to find beauty in our universe, our planet, and our lives as revealed by science, part of that must include shaking off the centuries of shame around sex. Not just for procreation but for fun.

It's a lucky thing there are lots of old and new ways to experience pleasure without anyone getting pregnant. The fact that sex and reproduction can be separate is almost as much of a miracle as the idea that one can cause the other. Countless people having great sex over the course of history were not in danger of reproducing, because they were the same sex, or postmenopausal, or just not doing it the way one does it in order to get pregnant. Or because some form of contraception, ancient and rudimentary or modern and nearly foolproof, was employed.

As I write this there are reports that new experiments in China have produced healthy baby mice from two moms. Maybe someday same-sex human couples will be able to meld into new people, but until then in vitro fertilization and other technologies make it possible for people who couldn't otherwise to have biological kids.

There are theologies that insist every orgasm must be in the service of conception, but in nature other species also have same-sex relationships, oral sex, and masturbation, and some even have sex when the female is already pregnant and therefore not able to conceive. (Our fellow primates, the bonobos, in particular.) There is clearly something wonderful about sex besides the chance to pass on your DNA.

I had a huge crush on Jon for a long time before anything romantic happened between us. For most of that time I had a different boyfriend who I was steadily falling out of love with. Jon wouldn't even signal that he was interested in me while I was still involved with the other guy. At the time, I couldn't tell if this was because he was a really good person or because he wasn't the least bit tempted. In either case, I didn't think we would date, let alone mix our genes to create a new human. I just hoped we would sleep together and best-case scenario could awkwardly still be friends. Once I was officially available we went with a big group of friends to a Mets game. It was a gorgeous hot summer night. I was wearing very short denim shorts. Have you ever seen a nature video where one animal, usually a male, maybe a rare brightly colored bird, is really putting his heart and soul into the mating ritual and the female is just so indifferent? And David Attenborough feels sorry for him because he's so desperate and pathetic? At Shea Stadium in the summer of 2005, I was that bird.

Believe me when I tell you I could not have come on any stronger.

Later that night, we shared a yellow cab downtown. At Twenty-Seventh Street, Jon gave me the most platonic hug, said something like "See you around," and got out. When the taxi door shut, I thought to myself, *Okay, I just have to accept that whatever he's into is not this.*

I had lost faith. But twenty-four hours later, after the clearing up of misunderstandings and the consuming of a large quantity of alcohol, we were having sex until dawn. It was fantastic. Attenborough would have been delighted for me, I'm sure.

For much of human history the elaborate ritual leading up to having sex with someone for the first time was a little different. It was a wedding.

For generations a woman's wedding day may have been full of firsts: the day she moved out of her parents' home, her introduction to her husband, and the day she lost her virginity. So much change all at once. Things are often different now. When I got married I didn't even change my name.

The honeymoon is left over from that world, a vestigial tradition for wading out of virginity. It's such a nice word. The etymology comes from the idea that the first month of marriage is sweet, like the honey mead supposedly drunk to aid with conception. Drawing on the full cycle of the moon, the first honeymoons were the initial heady month of married life, also the first ovulation cycle and chance to reproduce. Honeymoons were originally about seclusion, a kind of staycation. They offered time to get to know each other, to snuggle and try out sex, theoretically for the first time. And most important, to get pregnant. The first mention of this dates back to the mid-1500s. Almost three hundred years would go by before rich Brits started making it into a fancy vacation.

I loved our honeymoon. It was an extravagant gift from my mother, a trip of a lifetime to Africa. Being in new places together bonded us. So did the delight of every flight attendant, waiter, and concierge who heard we were just married. We were happy and they were happy for us. But there was also an implicated approval that it was a good thing to get married, a sense that commitment and monogamy are accomplishments.

They are, of course, not the only way to do things. A lot of human beings are monogamous, but definitely not everyone. For millennia, polygamy was extremely popular. It's still common in many places. Our hunter-gatherer ancestors may have practiced something akin to what we now might call group marriage. People have all sorts of systems, from celibacy to swinging. And why not? I don't believe that there are limits to the ways people can be happy. Different species of animals have different ways of organizing their mating. Prairie voles are monogamous. Elephant seals are polygamous. Chimps are promiscuous. When we study these species we don't impose moral judgments on them, we just seek to understand.

Once, at a dinner party, I happened to mention that I didn't think polygamy should be illegal because consenting adults should be allowed to arrange their families however they please. Someone objected, "So you're saying you'd like Jon to take another wife?!" No. For the record, I would definitely *not* like that. But I also understand that what is right for us is not right for everyone. The only litmus test, so far as I can tell, should be *Is anyone being hurt?* If there is no deceit, manipulation, or abuse involved, who are we to judge? And yet people get very worked up about these things, even in the abstract, even when it has nothing to do with them.

Why? Of all the things we do to keep our species going, why is it that sex evokes such strong opinions from people not directly involved in a given situation?

Eating, breathing, farming, gathering, building shelters: these are actions we take to try to keep going a little longer, to plan for the future. Even hunting, which is dangerous, gruesome, and violent, has historically had no shame or secrecy attached to it. Why is sex different? Sexual mores and norms have varied among communities, but at some point sex became widely considered one of the most taboo elements of life, at least in the West. How did this happen? Even the Bible acknowledged Adam and Eve had to "learn" to be embarrassed about just hanging out stark-naked. They weren't shy about it until the whole fruit-eating incident. It's from the Tree of Knowledge that Eve eats the forbidden fruit and ruins everything. In the language of the Bible, to "know" someone is a favorite euphemism for sex.

But sex is useful. Not just in terms of making babies but in terms of relieving stress, bonding you to someone you love, and helping you fall asleep. And it's extremely powerful. Not just because there is sometimes a chance at creating a life, but because the mere hope of it can make us do and say things we never would otherwise. It's like being put under a spell.

Maybe if sex wasn't so powerful we wouldn't burden it with so many rules, so many societal expectations. Maybe this explains our compulsion to try to tame it. Like everything enormous, wondrous, and scary in nature, it evokes both glee and anxiety. Like fire, like weather, like everything wild, the intensity of nature evokes fear in us. But we're part of nature. And sex is one moment in which we can almost grasp that.

To borrow a phrase: With great power comes great responsibility. Which may explain why there are detailed sets of norms and taboos about sex that members of any society must learn, some fair and logical, some prejudiced and pointless. Shame has been a pretty effective tool to keep people from breaking the rules, or at least to keep whatever they were doing under wraps. But the shame is not innate.

Professor Stephen Greenblatt argues that St. Augustine, the very influential fourth- and fifth-century Roman–North African Catholic theologian, invented sex, or rather our modern view of it. Regarding Adam and Eve, Greenblatt writes, "Pagans ridiculed that story as primitive and ethically incoherent. How could a god worthy of respect try to keep humans from the knowledge of good and evil? Jews and Christians of any sophistication preferred not to dwell upon it or distanced themselves by treating it as an allegory."

Greenblatt's position is that "the archaic story of the naked man and woman, the talking snake, and the magical trees was something of an embarrassment. It was Augustine who rescued it from the decorous oblivion to which it seemed to be heading. He bears principal responsibility for its prominence, including the fact that four in ten Americans today profess to believe in its literal truth. During the more than forty years that succeeded his momentous conversion—years of endless controversy and the wielding of power and feverish writing—he persuaded himself that it was no mere fable or myth. It was the key to everything."

The connection between lust and loss of control haunted and worried Augustine. These themes crescendoed in the form of an erection. Other body parts are easily controllable, but the penis has a mind of its own, and he did not like that. Greenblatt believes that this was primarily due to the deep shame Augustine felt when his own erection was pointed out by his father at a public bath. This set into motion a family drama that may have come to define him, and possibly Western civilization.

It's a reminder that biology is extremely powerful. Augustine figured this sense of being possessed must be somehow a little evil. What Augustine didn't understand is that it's the loss of control that makes it miraculous. Like so many other transcendent experiences, it's the giving over of control to nature that makes it sacred. We are powerless over so many of the most thrilling elements of life: the changing of the seasons, sunrises and sunsets, falling in love. These offer us a taste of the grandeur of being alive, a

reminder that we are part of nature. How can we not celebrate this?

Biologically, sex creates physical intoxicants inside our
bodies. Our brains dispense a shot of dopamine that makes
us feel euphoric, and then a dose of oxytocin, which makes
us feel bonded. That's where we get the ecstasy of falling in
love. We get high on our own supplies. These things can be
measured and tested. But they are still transcendent. Science gives us the ability to see how nature equipped us for
such joys.

But let's try to take a step back.

Sex can also lead to a range of problems. There are diseases. Childbirth can be fatal. Unplanned children are life
altering. Sex is fun because if it weren't, we wouldn't bother.
And species that don't like reproducing don't last.

So the biological urge must be stronger than the taboo.
It's not sustainable to abolish sex completely—although
some have tried. Just ask the Shakers. The United Society of
Believers in Christ's Second Appearing was an eighteenth-
century sect of Christians who believed in pacifism, com-
munal living, and abstinence. Not abstinence before
marriage. Total abstinence. They believed that the real original sin in the Garden of Eden was sex; it was a tree of that
kind of "knowledge." The sect's founder, Ann Lee, taught
that a life short of total celibacy was a life of impurity and

ungodliness. They believed conversion and adoption would be enough to keep their order alive. It was not.

Instead, most belief systems argue that sex is acceptable when it's for procreation, not recreation. Would that not be equivalent to only eating for nutrition? Empty calories might be frowned upon if you're trying to keep trim, but they are rarely considered a sin.

Sex as procreation and within the confines of marriage is frequently the stance supported by religious groups. Having many children is considered a way to honor God—by Catholics, Orthodox Jews, Mormons, and certain sects of evangelical Christians, among others. Many sacred teachings encourage followers to "be fruitful and multiply" in one way or another. It's a biological advantage. The more children you have, the better chances your genes have to live on long after your death.

However, some societies, large and small, have gone beyond seeing sex as a means to an end, and have also encouraged it more generally, incorporating it into their practice and celebrating it openly, treating it as holy.

The Oneida community, in upstate New York, was not part of the Oneida Native American tribe for which their town was named (although they sometimes did socialize with them) but were in fact a group of white people who were having a ton of sex. At their height they numbered about three hundred in the mid- to late-nineteenth century.

They invented household time-savers like mousetraps and lazy Susans, made silverware, and avoided monogamy at all costs. As Louis J. Kern wrote in his 1981 book, *An Ordered Love: Sex Roles and Sexuality in Victorian Utopians,* "the sexual act had all the significance of a sacrament" among the Oneida. They believed that monogamy was a form of selfishness, akin to other forms of social inequity like economic injustice and slavery, and that sharing (even sexual partners) meant caring. Just as there was no private property in their community, there were no private marriages.

In Oceania, a kind of ritualized swinging or wife-swapping custom was practiced in hopes of everything from expanding the gene pool to solidifying a pact. In the first decade of the twentieth century Van Gennep wrote in *The Rites of Passage* that "among the central Australian the sexual act is auxiliary to magic and is not a fertility rite." The idea being that even if you don't produce a child, sex is still wondrous, and meaningful as a kind of communication. Before the Incas, the Moche people inhabited what we now call Peru. They appear to have immortalized their religious sexual rites in their art, specifically pottery.

Even in our modern world, sex-positive communities thrive. Many are new, but some are not. For centuries, the animist Muria people, now a society of around seven thousand in the state of Chhattisgarh, India, have provided something like dorms for teens, called *ghotuls.* There is enormous social pressure to live there at least for a time before you marry. When there, young people learn skills, make art and pyres, and perform other tasks critical to the community's wedding and funeral rituals. And over the centuries it also became a place to pair up for an evening of lovemaking. When night falls it turns into a kind of rustic nightclub, complete with tobacco, drumming, and sensual massage. Nowadays there are even taboos against getting too attached to one person right away and sleeping with only them. Orgies, however, are not allowed.

But that restriction hasn't always applied to ritual sex, even among some so-called mainstream monotheistic religions. In her book *The Great Transformation*, Karen Armstrong describes the religious orgies of the Israelites, who at that time worshipped Baal, an early prototype and rival of the God familiar to modern Jews, Christians, and Muslims. Despite the disapproval of their prophets, group sex was a way to honor and harness the power of sex into an appeal for something they needed: a fruitful harvest. This practice continued "well into the eighth century and beyond."

Examples of *hieros gamos*, or holy marriage (ritual reenactments of a mythical sexual encounter between deities now performed by mortals), can be found from the ancient Greeks to Tantric Buddhists. Like sex scenes in mainstream movies, these were sometimes just dramatizations, but sometimes, as with another kind of movie, they were real sex acts performed as a way to get closer to the sacred.

Societies do not agree on the circumstances under which one should or should not be allowed to have sex, but there does always seem to be a rigid set of expectations.

In Amy Heckerling's 1995 teen classic, *Clueless*, her heroine Cher Horowitz's friend ruthlessly insults her by saying, "Why am I even listening to you to begin with? You're a virgin who can't drive." Audiences around the nation felt this scorn. Cher is humiliated. *Clueless* is adapted, albeit loosely, from Jane Austen's even more classic novel *Emma*. In the United Kingdom in 1815, when it was first published, being a virgin who can't drive would be very much expected from any unmarried teen. Sex was only for after marriage (and what's "driving"?). Had Emma Woodhouse had sex, she would have become a pariah, unfit for marriage and therefore worthless, but 180 years later sex would have transformed Cher from naive to wise, from someone unworthy of advice-giving to a source of potential insight.

Both young women faced societal expectations. The question was never just whether or not they felt personally inclined to have sex. For something that is so often described as "personal," the entire community, pro or con, reliably seems to feel they have a say. And the same rules often don't apply to men and women. Modern women in secular societies are still regularly shamed, privately and publicly, for having lots of partners or just a lot of enthusiasm about sex. For men in those same societies the social pressure goes the other way. Too little sexual experience can be the cause of ridicule. In both cases the idea is that there is a standard to be met, regardless of your actual desires.

No matter what the social mores of the time and place, sex, especially for the first time or with a new partner, is a portal from one reality to another. Everything can change: your relationship, your status in your society, sometimes your whole life. The night I finally slept with Jon proved to be a portal from friendship to romance. I still get a giddy thrill in the pit of my stomach when I think of it, despite the fact that it was thousands and thousands of nights ago. It took many years to arrive at the moment when another incarnation of the same ritual would, in time, transform us into parents. Through this sacred rite we were able to conjure a new person into existence. This is a story worthy of an ancient creation myth. And it's true.

chapter twelve

A Monthly Ritual

Do not swear by the moon, for she changes constantly.
—William Shakespeare, Romeo and Juliet

I remember a man who said a period was like shit. Well, sacred shit, I told him, because you wouldn't be here if periods didn't happen.
—Chimamanda Ngozi Adichie, Dear Ijeawele

For years I wondered why I didn't hear more about how the moon controls our menstrual cycles. Each a monthly process, surely they were connected. It seemed so beautiful, so poetic, such a special thing to be able to tell my daughter someday: *Soon your body will link up to that big rock that orbits us 238,900 miles up in the sky and then you will be able to create life.* Another story worthy of mythology. I couldn't understand why everyone didn't talk about it.

But in my research I discovered, to my astonishment,

that there is no scientific proof of a correlation between the phases of the moon and the human female menstrual cycle, even though they are both around twenty-eight days long. I don't think I was the first person ever to make this mistake. Artemis seems to be evidence of that. But that doesn't mean I hadn't still been completely and totally wrong. I don't know if someone told me this theory at some point, or if I made a misguided assumption, but either way I was utterly sure of something for decades that was not true. Even worse, I told other people, for which I am especially ashamed.

I had been intoxicated, again, by my human need to draw connections, to recognize patterns.

The connection between the cycle lengths appears to be pure coincidence. That isn't to say a link could never be discovered, but as of now, we don't have evidence of one. I'm disappointed, but still I crave to understand what's really going on, even if it's not the pretty myth I'd hoped for.

No matter what, the moon is still awe-inspiring, still central to the human experience. Think of how much the moon used to shape daily life, or rather nightly life. Before electricity, imagine the difference between an evening stroll under a full moon and a crescent. All over the planet, so many generations—speaking different languages, believing in different deities—looked up at the same shining body in the night sky. The Romans, the Muslims, the Jews, the Chinese, the Igbo and Yoruba of Nigeria, and the Haida of British

Columbia, among others, all more or less built their calendar around the lunar cycle. The moon, like so much in nature, provides a rhythm to time. That's why our months are the length they are. It's also why the words *month* and *moon* are so close.

The waxing and waning are so steady, clear, and reliable. Once humans understood the way the moon grows and shrinks, we could mathematically predict its movements perfectly. The moon is more orthodox than even the most devout zealot.

And there is another important rhythm on Earth that *is* orchestrated by the moon: the tides. Lots of clever people who lived in all manner of coastal civilizations may have noticed this, but their names and theories are lost to the ages. A Greek explorer called Pytheas, who lived during the fourth century BCE, was the first person, as far as we know, to deduce a connection between the lunar phases and the sea—but not why or how. That part took almost another two millennia, until Isaac Newton grasped the concept of gravity and figured out the pull the moon had on our oceans.

Until then, how easy it must have been to believe the connection was supernatural, the work of magic or of a god. Or rather goddess, as the moon has so often been personified as female. In poetry and myth we've been coupling our moon and our ideas of femininity for eons. The Shinto sun goddess has a lunar counterpart called Tsukiyomi-no-mikoto, who is the spiritual incarnation of the domain of darkness. In Hawaii, the moon goddess Lona married an Earthling. For the Fon of Benin, Gleti is the moon goddess and mother to the stars. Changxi, the ancient Chinese version, is sister-wife to the sun goddess Xihe. In ancient Rome she was called Luna, and she's still with us in words like *lunar*. From the Maya to the Shawnee, from the indigenous people of the Philippines to the indigenous people of Finland, the moon has been considered both overtly womanly and worthy of worship.

Perhaps the connection comes from the parallel natural cycle of our bodies. Or maybe it's because, for so much of human history, women have been allowed so few roles besides reflecting the light of their fathers, husbands, and sons, the way the moon reflects the light of our star.

Despite the fact that our species depends on it, menstruation has often been considered taboo, impure, and hazardous. All over the world, men appear to have been scared by it and have taken great pains to avoid coming in contact with it. So much so that women from Nepal to Nigeria to the Yurok tribe of North America have been traditionally exiled into huts or other structures during their periods. This segregation is sometimes dangerous: women still die in this isolation from unsafe conditions. This is, of course, the extreme. But there are all sorts of special rules for people who happen to be shedding their uterine lining—notably among some traditional Jews and Muslims, some sects of Rastafarians and Buddhists, some Aboriginals, and some Eastern Orthodox Christians. Temporary abstinence is a popular one. For some it's no cooking. For others it's no visiting the place of worship. And when it's over, there are ritual cleanings. For Orthodox Jewish women, the post-menstrual ritual bath is called a mikvah and is required before resuming sex. Because of the timing of our cycles, these rituals are monthly.

What a different world this would be if, instead of prohibitions and sexual bylaws, we had celebrations when our periods came. What if there were special things we get to do instead of things we cannot?

One of the things most deserving of celebration in my life has been my immense good fortune in friends. I have very dear, lifelong male friends who are like brothers to me and to whom I owe a lot, but my bonds with my girlfriends make up my daily life. In person, on the phone, over email and text, we are together through the hills and valleys. They are a constant source of strength and joy for me. I hope I am for them as well. I crave a community of women. This is something I might have had if I were religious, a perk of being active at temple or mosque or church. So many religious sects have women's groups, for prayer or study, fund-raising or charity, or just to join together to do whatever it is that is expected of us. It's one of the hardest parts of being secular: you have to work to congregate.

In 2009, Jon and I returned to New York after living in London for two years. I was overwhelmed by the desire to have a thousand long dinners with my friends, each of whom I had missed terribly while abroad. I wanted to cherish them individually, hear every anecdote I had missed. And when, months later, a close girlfriend from London came to visit, I invited a whole gaggle of my most beloved ladies to a girls-only dinner.

Looking around the table, I realized my earlier system of seeing these women mostly one or two at a time was absurd. I had an enormous wealth of spectacular women in my life and I was selfishly hoarding them. They should know one another. Not just casually, not just on social media, not just to say hello, but truly and deeply know one another.

In the autumn of 2010 the idea of a monthly dinner came to me. I chose a restaurant, centrally located near Union Square's rainbow of subway lines, reasonably priced but still special, with a fun cocktail menu and a flexible reservations policy. It also had a photo booth, which somehow seemed crucial.

In a fanciful spasm I decided the ritual I was attempting to create should be called the Ladies Dining Society. I sent out an email to eighteen of the coolest women I knew from childhood, college, work, and life.

Six showed up. Including me.

I was nervous. Would there be enough to talk about? Was this too contrived? Too silly? But the initial awkwardness melted away. And as we drank Dirty Janes, martinis garnished with pickled green tomatoes (soon a staple of our dinners), I watched the women I so admired start to fall in love with one another.

More ladies came the next month. Some of them were my lifelong besties from Ithaca, some I had met only recently. Filmmakers, teachers, writers, therapists, designers, executives, actresses, waitresses, resource managers, and stay-at-home moms. Among them, they have been all over the planet and experienced so many ways to be a woman on Earth.

And soon it felt like a real tradition. At each dinner we dined and drank and discussed politics and movies and mothers and motherhood and love and sex and art and everything else that makes the world interesting. We left rejuvenated if not a little drunk. Some women came almost every month, others only once. As I met new women, I invited them in. Our group grew with new perspectives and ideas.

As time went on, my initial intent came to fruition. Deep friendships bloomed. I had wanted to create a community and here it was, taking shape. One day, two of the ladies showed up together with bags from my favorite lingerie shop.

"We went bra shopping together!"

I was gratified; two women I loved were now independently friends. But I can't deny I was also jealous. *You two are friends now. I am not central to this equation.* This ritual had been giving me a sense of certainty and now it was evolving, taking on a life of its own without me.

As time passed I created a kind of dogma for myself. There were complaints once when I had the dinner on a weekend. After that it was strictly a weeknight affair. I'd make the reservation for five people, then send out the initial email and extend it once I knew how many ladies would be attending. Then two days before I would send out a reminder email. I kept the style of the emails uniform, effusive but formal, always signing them the same way. I would make the reservation for 7:30 p.m. but alert the hostess we'd likely not sit until 8:15, making for a little cocktail hour at the bar as everyone trickled in.

I would keep track of it all—who was invited, who came, who didn't—on an enormous spreadsheet that eventually spanned years. I even kept track of what I wore so as not to repeat the same dress too often. I felt a sense of anxiety about making everything uniform, perfect, and consistent. It gave me the illusion of control.

Yet I couldn't ignore something that was wholly out of my control, no matter how pristine I kept my spreadsheet. There was a slow but steady trickle away from New York. It's a hard place in which to settle down forever. It can be done, but other towns are easier. Work or love called so many of my friends elsewhere, often westward to California. But no matter how far away they moved I always kept them on the email list, hoping by chance a meeting or a vacation might bring them through New York on the night of our ritual.

As my friends came and went, the tradition itself remained steadfast. For almost five years, we missed only three monthly dinners: the two before my wedding, because I was too consumed with planning, and the one in October 2012, when Hurricane Sandy left lower Manhattan in the cold, wet dark.

Over the course of that time, babies were born, family members died. We lived one another's romances and breakups. We started new jobs and new careers. We got fired or quit. We fell down. We got back up. We shared the milestones of our lives.

And that's how the Ladies Dining Society went along, until April of 2015, when Jon got a job offer that was too good to pass up—with one catch. We'd have to move to Boston. Soon, I too was called away from my most beloved city, my friends, the Dirty Janes, and our ritual.

Around the same time, in LA, one of my lifelong Ithaca girlfriends had the idea she wanted to start a West Coast chapter of the Ladies Dining Society. Over the phone, I gave her specific, zealous, rigid instructions, as if for baking an impossibly fussy soufflé: "Put 7:30 p.m. on the email, but allow for forty-five minutes of cocktail time . . ." "Invite women who don't know each other . . ." "Have it on a week-night!" "Find a place with a photo booth!"

She listened carefully, enthusiastically, but a few weeks later she reported back that the Los Angeles Ladies Dining Society was different from the New York incarnation. The ladies already knew one another; many were members of the same theater company. Sunday nights were most conve-nient for them. Great cocktails weren't a real concern—people are so health-conscious in LA, and people have to drive in LA. There was no photo booth.

But it was nonetheless magical. The ladies had all had a ball and wanted to do it again the next month.

The tradition had taken on a life of its own. It had mu-tated. It had to, in order to continue. The orthodoxy of the New York Ladies Dining Society fit the needs of our sect, but other sects in other places had different ways of cele-brating their bonds of sisterhood. I suppressed my urge to be fanatical, and felt grateful something we started was out in the world, thriving.

In his 2006 book *Breaking the Spell: Religion as a Natural Phenomenon*, about how science can be used to examine faith, Daniel C. Dennett wrote, "In the face of inevitable wear and tear, no designed thing persists for long without renewal and replication. The institutions and habits of hu-man culture are just as bound by this principle, the second law of thermodynamics, as are the organisms, organs and instincts of biology." This is kind of what happened with the Ladies Dining Society.

Two more friends who have also left New York have since created their own versions of this ritual. New to Oklahoma City, my first college friend started a Ladies Dining Society devoted to trying new restaurants and seeing classic movies. In Los Angeles, a male friend has started the Brothers Dining Society for gay men of color in show biz. I have spared them both the list of directions.

As we settled into life in Boston I started getting on the Amtrak less frequently. I eventually made local friends and even more of my girlfriends left New York. The idea of schlepping down there with Helena and all the accoutrements a baby requires felt overwhelming. I've let the tradition I worked hard to build slip away from me. Circumstances change and people adapt. This is the way of so many customs, large and small.

But just as easily as these traditions fall away, new ones can begin. There's nothing to stop you from writing to your friends today and trying something new. A book club or movie night, a poker game or a cocktail party. Something steady, something special, just twelve times a year. You'll have to decide what the parameters are for yourself. Some elements might work out. Some will not. Trial and error will be the only way to do this kind of experiment.

Science has its own orthodoxy. If an idea or a theory cannot stand up to scrutiny, we must let it go. Testable proof and hard evidence dictate what is real and what is not. And like all orthodox people, I sometimes fail to meet the standards I claim to live by. No matter how much I'd like to, I cannot produce evidence that the moon controls our menstrual cycles. Maybe some future Newton will come along and enlighten us. Maybe not. Until then I still relish how the moon does have certain powers over us. It dictates our calendar and, the world over, inspires us to imagine goddesses. And every month, at least for a while, I got to dine with a pantheon of real ones.

*

chapter thirteen

Autumn

Twinkle those small stars,
In Orion, in the Pleiades.
Shrinking, through the dark we walk . . .
  —Shijing *or* The Book of Songs

A s with spring, our idea of autumn has a few compo-
nents. The fall equinox signals the shortening of the
days. The weather gets colder. The leaves start to change
color and drop from their branches. Food is harvested.
Plants wither. Animals prepare for hibernation. These her-
ald a thematic shift and remind us: *Darkness and death are*
coming. They're not here yet, but they're on their way, so live
while you still can.

This narrative is apparent in harvest festivals and in the
variety of holidays we've created to find joy in our terror,
glee in the ghoulish, freeing ourselves to a kind of morbid

fun. In autumn we wrap ourselves in the coming darkness, taking death's power and making it our own.

As an adult, I love Halloween. I like the break from societal norms. I like the Dionysian frenzy. I like the window into the fantasy alter egos of strangers. I like the freedom. In my twenties especially, I appreciated the opportunity to dress in provocative, probably very flammable nurse or fairy-tale costumes, free from judgment, if only for one night.

But when I was small, I dreaded Halloween all year long. It wasn't the cotton cobwebs or the haunted houses or the showing up unannounced at the homes of strangers. It was a custom that is, as far as I can tell, unique to my otherwise really lovely and not-at-all psychologically traumatizing elementary school. For reasons still not entirely clear to me, every Halloween, all my costumed schoolmates and I were marched five blocks to a nearby assisted living facility to participate in a parade for the residents. In a line of what must have been a few hundred tiny vampires, ghosts, fairies, pirates, and Teenage Mutant Ninja Turtles, we'd wind in and out of common rooms, up and down stairwells.

I have no doubt that seeing little children in their Halloween best was a much-needed source of joy for the residents. And many of them were cheerful, friendly grandparents who were in no way scary. There was, however, a large contingent of residents who were very clearly near the end of their lives. Through my little eyes I saw them: toothless, gray, skeletal, unable to speak, moaning, reaching for us or just frozen by age. I remember them as being as close to the living dead as anything I have seen before or since. Even then I understood that this was not their fault, they weren't trying to scare us, but it was impossible not to feel like we were facing the stuff of nightmares. These people were petrifying, both literally and figuratively.

I don't remember being told beforehand that this would be scary, or any teacher or aide acknowledging that this was, at best, a double-edged sword. In adulthood, my life-long girlfriends and I finally admitted to one another how upsetting this was and how much we would have benefited from some warning, or tools to process what we were see-ing. I think, being small and unsure of all the ins and outs of my own society, I thought this was just part of Halloween. We scare them with our felt and poster board costumes, and they scare us with their palpable mortality. Not because they wanted to terrorize us, of course, but because this is where they are in the cycle of their lives.

In middle school it became clear that other children did not take part in this ritual. I realized that the dress-up, the candy, and the low-grade set design were the "real" Halloween. I was annoyed. I thought I had been robbed of the true meaning of this holiday. But only because I didn't understand its origins.

Even on the clearest night, in the most remote, unpolluted corner of this world, we can only see a tiny fraction of all the stars in our universe. But those we can see have had an enormous impact on the lives of us Earthlings—although not in the way that the writers of magazine horoscopes would have you believe. There is no information up there about whether or not your crush will call, but there's another way in which the stars affect us. They have, since the dawn of time, had a large and real influence over our imaginations and our calendars. Disparate societies in far-flung corners of the planet have often, without any ability to communicate with one another, based their most beloved stories around the same bright corners of the night sky.

In one particular corner you'll find seven hot blue stars, closely clustered.

They are more than four hundred light-years away. (That is to say, if we could somehow manage to travel at the absolute speed limit allowed by the physical laws of the universe, it would still take us something like twenty generations to get there from the point of view of those left on Earth. Those traveling would lose almost no time at all thanks to time dilation.) Yet even from this vast distance, these particular stars have managed to weave themselves into the lives of human beings across the planet.

The Cayuga, the people from whom my hometown was taken, waited until the stars were overhead in winter to celebrate their new year with a seven-day dream-analyzing festival that would have made Freud proud—a kind of celebration of the subconscious. The Tuareg people of the North African desert used these stars for centuries to predict the changing of the seasons. The ancient Turks saw the cluster as a small battalion in formation. In South Africa, the Khoikhoi tribe saw them as the omen that the rains are coming. Not far away, the Xhosa called them Isilimela, or the "digging stars," the signal to start tilling the earth. The Aztecs spoke of these stars in Nahuatl as "the marketplace." North of them, the Mono people, who lived in California centuries before it was California, mythologized these stars as a group of ladies who preferred onions to husbands. The ancient Greeks saw seven sisters—a cadre of nannies for a baby Dionysus. In Japan, they are seen as a merging of several into one, a *subaru*, for which sensible midrange station wagons and flagship telescopes have been named. In English we call them the Pleiades, or the Seven Sisters, a group of bright blue stars that came to be about 100 million years ago.

In Quechua, Maruja's first language, the language of the pre-Colombian people of the Andes Mountains, the stars were "the storehouse." In the Andes Mountains, for centuries before Maruja was a nun there, her ancestors celebrated Quyllurit'i when the Pleiades reappeared after spending two months out of view. Quyllurit'i means "bright white snow" and falls in late May or early June, just before the winter solstice for the people of the southern hemisphere. It's a celebration of this light as darkness approaches.

But it's the Celts, the Iron Age and medieval tenants of what is currently known as the British Isles, who connected the Pleiades and their movement across the sky to the harvest, the dying of the light, and our relationship with our own death. Their festival of Samhain is the great-grandmother of our Halloween. It was celebrated when the Pleiades reached the highest point in the night sky. This coincided with the moment in the year exactly between the autumnal equinox and the winter solstice. On Samhain, pre-Christian Brits would look at their cattle, their pigs, and their sheep and decide which would survive and which would be devoured during the coming winter. On the highest hills bonfires were lit, feasts were served, and the souls of the dead were invited to join the party.

Halloween may be the great-grandchild of Samhain, but it's not its only living relative. For the planet's 1.2 billion Catholics there is All Souls' Day, a time to honor and pray for the dearly departed and, originally, to help those stuck in Purgatory graduate to Heaven. In modern Mexico, October 31 to November 2 is Día de los Muertos, the Day of the Dead, a Technicolor celebration of mortality where the spirits of dead children are invited back for a day and graves are decorated by those who miss them most.

We make rituals out of facing our fears, wrestling with them, even making light of them, beautifying them. Life used to be scarier. We evolved from creatures who were prey in the darkness of a world where we had not yet gained control of fire. No light but the moon and the stars. Imagine those long winter nights. Imagine what an advantage fear was and how quickly those who were unsuspicious of the sounds in the forest were devoured.

Now, many of us (but it must be said, certainly, tragically, nowhere near all) sleep safely, bellies full, soundly locked in houses easily flooded with light at the flip of a switch. But the fear is still inside us. We let it out on roller coasters, at horror movies, and, in many cultures, through rituals as the nights get long and dark.

How can something as old as lighting candles remain so popular? Whether it's a prayer candle or the one inside a jack-o'-lantern, we find a way to incorporate this small magic trick in so many rituals. We must be still astonished that we managed to create a little day in the night. We've had it in our celebratory arsenal for more than a million years and the novelty has yet to wear off or go out of style.

Dewali is the five-day-long Hindu festival of lights, which takes place in October or November each year. For some Hindus it's about the deity Rama's return from exile. For others it honors Lakshmi, the goddess of riches. Jains and Sikhs celebrate Dewali as well, but back it with different legends. Across the board, the ritual lamp-lighting symbolizes how our understanding drives out darkness. It's an autumn holiday that originated in the northern hemisphere, but like summertime Christmas in the southern hemisphere, it's popular and widespread enough that in places like Fiji and Mauritius it is a spring national holiday.

Since the Heian period in medieval Japan, Tsukimi, the moon-viewing festival, has celebrated a different kind of light every autumn. Special foods are made as offerings to the moon, and eaten here on Earth as well. These include dumplings, sweet potato, and mooncakes. As the moon reflects the sun, this too is a celebration of a small flicker of light in the darkness, giving people hope to carry forward.

Elsewhere in Asia where the end of monsoon season comes in late August or early September, the climate cools and ancestors are honored. The Chinese celebrate the Ghost Festival, when it is believed that the dead visit the living. In September and October, Cambodians spend the fifteen straight days of Pchum Ben honoring their ancestors when the gates of hell supposedly open to unleash generations of hungry ghosts. In Vietnam the equivalent holiday is Tet Trung Nguyen, when the poor should be fed, birds and fish released back into the wild, and the souls of the damned liberated from hell.

These rituals all serve as reminders that death, like winter, is always coming. They are inextricably linked and there is no escape from either. Halloween warns us that there is something powerful and mysterious coming for each of us, and that before it does we must relish the present with glee.

It is that dichotomy, between our present bright, sharp aliveness and our inevitable, approaching end, that's at the heart of all our fears, the root of all our wishful thinking, and in all our worst nightmares: in the nursing home in downtown Ithaca, at every haunted hayride and Halloween parade, in every ghost story that chills us.

I don't know if my elementary school teachers intended this lesson, but the local assisted living facility was actually the ideal place to experience the truest roots of Halloween and its many holiday cousins. It's easy to ignore this darkness, especially in a modern world where we are largely removed from the realities of our dying, and of our own survival. But we should see it, know it, understand it so we can appreciate the light—be it candle, star, or moon.

Halloween has another element, too. It's a break from the rules of society. Day in and day out we abide by so many detailed customs and expectations. What we wear, what we say, what we do are dictated, at least in part, by social norms. But on Halloween, we can become another secret self in front of everyone with no consequences. It's a big loophole, really. Imagine showing up at a bar dressed as a vampire on some random evening in March. But we need the escape. And it must be a very deep need, because throughout history societies have created little valves to release the pressure they create. Without this the tension builds up and the whole system cannot hold. Around the world and throughout the ages, we have ritualistically masqueraded as all manner of animal and god, spirit and ancestor, bending gender and time.

The global, ancient art of mask making is clear evidence that we love to, at special times, take on the identity of someone or something else. For the Dogon of Mali, like many other sub-Saharan societies, there is a long history of this art, with a wide range of subjects. Included among them is the *kanaga* mask worn to help honor the dead and escort their souls out of the village. The Dan (or Gio) people of West Africa make miniature masks as sacred objects that can easily come with you when the regular masks cannot. For the adolescent boys of the western Pende of the Democratic Republic of the Congo, there are special feathered masks for sneaking up on people. Beyond Africa, there is ritual mask making from the Inuit, whose shamans carve masks to reflect their visions of the spirit world, to Korea, where ceremonial masks are as ancient as the society itself and still used to perform social satire, to the depictions of masked hunters in the cave paintings of prehistoric Europe.

On the second Halloween of Helena's life, I tried to dress her up as a lion. She roundly rejected the construction-paper paws, tail, nose, and whiskers I had made but was willing to don the store-bought mane and ears, albeit briefly. Jon was stuck working late, so she and I went for a long walk in the early evening. She was still too little for candy, so we didn't trick-or-treat, just went to look at everyone in their regalia, letting out their secret alter egos. We made our way to Beacon Hill to roam the narrow brick streets built atop cow paths dating back to the days of witch trials. We marveled at the brownstones decorated with spiderwebs and mummies. To me it was magic. "Look at everyone in their costumes! Isn't this wonderful?" I asked Helena, who by that point had removed her mane and was looking up at me from her stroller with an increasingly skeptical expression. Soon it was dark and the streets were filled with families squealing with laughter, high on sugar. I don't know if it was the crowds or the visceral sense that the rules had been turned upside down, but Helena's usual curious good nature was fading.

"Isn't Halloween exciting?" I asked.

She was still new to talking but responded clearly, shouting, "All done! All done!" employing a phrase we had taught her so she could signal when she was finished eating. I had never heard her use it in any other context before.

"Are you saying you're all done with Halloween?"

"All done! All done! All done! All done!"

"Okay, okay, we're going home."

Her chorus of "All done! All done! All done! All done!"

continued until we were well away from the costumed madness.

Maybe next year she'll like it more. Or Halloween might not be for her. As Helena grows, more will be revealed.

Two years before, I'd been home in Ithaca. I had just said goodbye to the closest thing I had had to a living father in almost twenty years. I watched as my grandpa Harry slipped into that limbo between life and death, which I'd first seen in the nursing home. The day he died, when I got my period, desperately wishing to be pregnant, the sun set earlier than it had the day before. The leaves were falling off the trees. Everything seemed to be dying. A vague, intangible feeling of hopelessness set in. I became very depressed. I was depressed because I didn't know that Jon and I would conceive Helena about a week later. I was depressed because I didn't know that by the time of the winter solstice I would be telling my mother the exciting news. I didn't know that two of my lifelong best girlfriends would soon tell me that they too were pregnant. I didn't know that a year from then we'd be together with our babies, our partners, and our entire group of best friends on one of the greatest vacations of our lives. I didn't know that on that vacation my girlfriends, Jon, and I would laugh and marvel at how much Helena reminded us of Grandpa Harry. I didn't know that I would see glimmers of his face, brief echoes of his expressions, the sounds of his sighs in my little girl, who carries him in one-eighth of her DNA.

I had forgotten that the light would, in fact, come back. And it was easy to wallow in the darkness.

Fourteen Octobers earlier, in 2002, I was visiting Rome during my semester abroad in Florence. My friends and I entered a small museum not far from the Spanish Steps, called the Capuchin Crypt. It's been there for centuries and I cannot imagine that a single person who has stepped inside has left unchanged. The crypt consists of six rooms ornately decorated with the human skeletal remains of a long line of members of this Christian brotherhood. A room of tibias, fibulas, and femurs. A room of sacra, ilia, and coccyges. A room of skulls. Thousands of human parts. Just like the ones inside each of us right now.

And in the last room there is a sign that reads, in several languages: *What you are now we used to be; what we are now you will be . . .*

I was just shy of twenty when I saw this, and scarcely a day goes by that I don't think of it. Secular Jew that I am, I would have never dreamed that one of the most profoundly stirring spiritual messages I would ever encounter would be found in a Catholic crypt in Rome. But that was due to my own narrow-mindedness. Each of us, no matter our beliefs or lack thereof, is wrestling with the deep knowledge that whatever comes next, *this*, what we are experiencing at this moment, will end with total finality. Whether we find nothingness or somethingness, it will be new and different from existence as we know it. For even if you believe in, say, reincarnation, life at another time in another body would be profoundly different from what you are experiencing today. Whatever is next, it comes for all of us.

Even the seven stars of the Pleiades will die someday. Instead of putting it out of our minds, instead of ignoring our fears, let's honor them, talk about them, and take pleasure in the light a little longer, before it's gone.

*

chapter fourteen

Feast & Fast

*[Fasting] is the cause of awakening man. The heart becomes
tender and the spirituality of man increases.*

—'ABDU'L BAHÁ

*For the Sun, as the universal father, sparks the principle of
growth in nature, and in the patient and fruitful womb of our
mother, the Earth, are hidden embryos of plants and men.*

—OHIYESA

My dad had been born into poverty. Not destitute but
poor. By contrast, his great-grandparents on his
mother Rachel's side starved to death around the time of
the Russian Revolution. A picture of them—her in her ba-
bushka shawl, head covered, him in the black coat of his
orthodoxy—hung in our house where we never wanted for
anything, specifically just off our kitchen, which was always
fully stocked. A reminder that we were lucky. Something
my parents made sure we understood.

I know that I have been extremely privileged. I got to see much of the world before I entered high school. And because of my parents' success they were able to bring Maruja into our home, who enriched my life with more love and knowledge than I would have had otherwise. I had all this and more not because I deserved it or had earned it in any way. I was just really lucky. Pure random chance. Like winning the lottery without even making the effort to buy a ticket.

I am profoundly grateful that I've never known what it is to go hungry, but for much of human history everyone knew what it was like. There were scarce times and plentiful times. A bad crop, a cold snap, a failed hunt, a long winter— all killed. It was part of life.

Middle-class Americans have access to more food and more varieties of food today than the most powerful rulers of just a few centuries ago. I doubt many would trade a trip to a local grocery store for the menu of a medieval king. We can get almost any fruit or vegetable at any time of the year, something unthinkable until very recently. But it's an easy thing to take for granted. There is an ancient, widespread ritual that can help us remember how lucky we are. For varying lengths of time and in varying frequency, almost every religion calls for a fast.

When I was growing up, we did not observe Yom Kippur, the Jewish Day of Atonement, but in my thirties I have adopted this one aspect of the most solemn, sober, holy day of my ancestors' year: fasting. For one day in September or October, depending on the year, from sunset to the following nightfall, I don't eat anything. This is a very reformed version of what Leviticus calls for. Here are some other things Leviticus says I'm not supposed to do on Yom Kippur: drink water, take a shower, moisturize, work, have sex, wear leather shoes. Those are a bridge too far for me. Here is something Leviticus says I *am* supposed to do on Yom Kippur: pray. I do not. Instead, I think about what it would be like if this was my everyday experience. I think about how many millions of people around the world and through the ages have considered this aching in the pit of their stomach normal. It's a kind of meditation, which might be another word for a secular prayer.

When Helena is bigger I plan to tell her why I'm fasting on Yom Kippur, to tell her how lucky we are to have food whenever we want it. And when she's big enough, she can make her own decision about joining me. Children are not expected to fast, of course. Nor are nursing mothers, but the autumn after I had Helena, I felt a deep, ancient compulsion to comply with the tradition. I thought of my great-grandfather Benjamin walking to temple while dying of stomach cancer. If he could do that every week, I could fast for one day even while breastfeeding. Maybe because we were now establishing family rituals, not just couple rituals, for the first time, Jon fasted with me. I found this so romantic. He is often very willing to participate in whatever traditions I am exploring or concocting, but celebrations and candle lightings are easy to get into. You don't have to give anything up. Fasting is a sacrifice, so in a way it meant more to me. We spent the day hungry, slightly cranky, and quiet, but united. Mostly I just remember being exhausted, but that could have been on account of the newborn.

Later, Jon said that more than anything the experience made him think about the "industrial food complex" and how the mass production of food was a kind of overcorrection to our deep, ancient fear of starvation. When we think about all the health problems in the United States that are due to obesity, it's like a multi-millennia-long *Twilight Zone* episode. We, our species, got what we so desperately longed for in all those lean times, but there's a catch. All this easy access, all these too-full bellies will kill us, too.

Fasting is, of course, not unique to Judaism. During Ramadan, which is based on the Islamic lunar calendar and therefore migrates throughout the widely used Gregorian year, Muslims fast every day from sunrise to sundown for a month. Besides introspection and stronger devotion to God, this ritual is designed to create empathy for those in need. Eid al-Fitr marks the end of Ramadan. The specifics vary from place to place, but across the Muslim world there are lavish feasts, charitable giving, fireworks, and moon viewing.

Leading up to Naw-Rúz, their springtime new year celebration, the Baha'i fast from sunrise to sunset for nineteen straight days, in the hopes of getting closer to God through withholding all the material things we desire but do not need.

Hindus fast regularly throughout the year, on holidays like Ekadashi, Navratri, Vijayadashami, or Karwa Chauth, as well as on set days of the week or month depending on their personal beliefs, region, local culture, and favorite gods. For example, a devotee of Krishna in the north of India may fast on a totally different schedule than someone who feels most connected to Ganesha in the south.

The Absaroke people of the American Great Plains (who were renamed Crow by the confused Europeans) couple religious fasting with religious dancing, in a ceremony still performed today. The combination of physical exertion and lack of food and water "humble" the dancers, bringing them closer to their cosmic vulnerability.

All year long, Mormons fast one Sunday a month, forgoing two meals and donating what they would have cost to the needy. There is absolutely nothing about this that requires belief in anything other than empathy. It's just the kind of thing any one of us could adopt to remind ourselves to be more charitable.

If you're open to it, you can find a religious reason to not eat almost any day of the year. But why do so many traditions include fasting? Why encourage followers not to eat when there's plenty? Maybe it's because our bodies have evolved for it. Whether it's somehow good for us or just tolerable, we have adapted to this because nature provides unsteadily, and species that cannot withstand lean times cannot survive the wait until the fat ones. In either case, religions that call for fasting may have unwittingly tapped into some deep biological programming.

But this isn't just an intellectual exercise. Hunger is real. Right now people are starving to death, like my great-great-grandparents did. In September of 2017, the World Health Organization reported that "after steadily declining for over a decade, global hunger is on the rise again, affecting 815 million people in 2016, or 11 per cent of the global population."

Not because there's not enough food on Earth, but because it is unevenly distributed.

We see the ads on TV, the faces of real human beings on the brink of death, including children, and find a way to separate ourselves from that reality. We pretend it's not our responsibility. I am guilty of this. How many times have I bought something that I do not really need while some other mother watches her child slip away?

With my chubby, healthy baby playing just a few feet away, I do not know how I could possibly justify my actions.

Subconsciously I convince myself we are different, me and that other mother, even though we are not. Maybe by fasting once in a while, maybe by meditating on the experience of all those who have a knot of hunger in their bellies every day, I can force my own hand and do more for others. I want this realization to fuel me to be more charitable, donate more, volunteer more, and appreciate better what I have. If more Americans understood the biological alarms of hunger, imagine what the response might be.

The other half of a fasting ritual is the chance to eat again, with a new appreciation for what you have. If the fast is a way of saying *sorry*, the breaking of the fast says *thank you*.

For about a billion years life-forms on Earth didn't eat other life-forms but subsisted on simple chemical energy. Slowly over the millennia Earthlings had to evolve the tools to break down food and make it into energy in new and complicated ways. It took a really long time. Now we do it with ease. It's the stuff of miracles.

The fat times deserve to be honored, too. Human beings celebrate harvest festivals the world over. A good harvest provides a collective sigh of relief. The fear of starvation is lifted, at least for the moment. The hard work has paid off. These are the joyful moments that bond communities, that give us the strength we need later, literally and figuratively.

It's not unusual for there to be only one belle at these

balls. Often a single crop carries much of the weight of sustaining a whole society, and the moment it blooms, ripens, or sprouts is a reason to celebrate, and show gratitude to a deity.

The Green Corn Ceremony is practiced at the beginning of harvest among Creek, Cherokee, Choctaw, Seminole, Haudenosaunee, and other Native American tribes. Besides celebrating the harvest of a plant crucial to survival, this festival features dance and ritual forgiveness. It takes place when the first hint of an ear of corn appears, kind of like Blossom Day and countless other "first fruit" festivals.

Rice was the cause for celebration for the Masaru ceremony among the Paiwan, who are the indigenous people of Taiwan, and for regional Shinto harvest festivals all over Japan. In Barbados, it was sugarcane. It's the New Yam Festival, the Iwa Ji, among the Igbo people of West Africa. In Cornwall, England, it was wheat, and in the Great Basin of the American West pine nuts were welcomed with dancing. La Fiesta Nacional de la Vendimia, a celebration of the moment when grapes are ready to transition into wine, has been marked in Argentina since the 1600s. These plant-human relationships deserve to be celebrated and they have been consistently and enthusiastically for eons. They are indisputably real and powerful. How grateful we should be to the plants. We eat them, or we eat animals that eat them.

Trees give us the oxygen we breathe. We shouldn't forget that we literally could not live without them.

Popul Vuh, the Maya creation myth, honors our connection to the plant world. It's the story of how our species was made from maize. And we are—Americans in particular. Today our diet is largely corn, disguised in millions of other foods. But no matter what we eat, the food we consume becomes a part of us. Our bodies are made up of cells that pull nutrients out of what we consume to function, to keep us alive.

It's so obvious, something we learn so early in childhood, that the stunning magic of this is easy to lose. But we shouldn't take it for granted. Something grows in the earth. Water, matter in the soil, and the light of our nearest star make it bigger. It changes color, blossoms, enlarges, transforms, appearing seemingly out of nowhere. We humans come along, take it, transform it with the semi-sacred task of cooking, and then we put it in our mouths, crunch it up with our teeth, pass it down our throats, and live on.

If I hadn't been aware of this all my life, I would consider it one of the greatest miracles of all. Something out of a fairy tale or science fiction.

Most people in the industrialized world are lucky enough to have some variety in our diet. Maybe you've eaten something different for dinner every night this week. Maybe you even ask yourself at seven p.m. what you're in the mood for and by seven thirty it arrives at your door. Anything you want. It doesn't even have to be in season. Can you imagine waiting all year for that first strawberry?

These relationships with plants are not as one-sided as they seem. We humans appear to have the upper hand, but as Yuval Noah Harari wrote in his book *Sapiens: A Brief History of Humankind*, "We did not domesticate wheat. It domesticated us. The word 'domesticate' comes from the Latin *domus*, which means 'house.' Who's the one living in a house? Not the wheat." Harari argues that this relationship called "agriculture" is problematic for humans, that we got the short end of the stick, unwittingly tricked by a plant into tirelessly farming, helping it multiply and flourish across the world. Would we be better off if we all just kept hunting and gathering?

A complex, tumultuous, ten-millennia-long multispecies love story, to borrow a phrase from my mother, and just a shadow of the romantic epic that is the entire ecosystem of the planet.

At our Thanksgivings, my mother raises a glass and says, "You don't need to know who to thank to be thankful!" Thanksgiving today, a kind of harvest festival, might be America's best-loved secular holiday, but officially, the modern incarnation of Thanksgiving was introduced as expressly theistic. After decades of unofficial celebration, President

Lincoln finally established it as a national holiday during the height of the Civil War, at the urging of magazine editor and "Mary Had a Little Lamb" composer Sarah Josepha Hale, in hopes of unifying the bitterly divided country. His proclamation specifies who should be thanked on Thanksgiving: "Most High God."

But even in its secular incarnation, Thanksgiving is built on lore that does not jibe with the historical facts. The idea of a peaceful breaking of bread where the Pilgrims had nothing but good intentions toward their Native American hosts is a national mythology, not a religious one. Whatever took place between the Wampanoag and the European settlers in Plymouth, it was not the inclusive dinner party we're told about in school. There's no evidence a single Native American was present. The only certainty is that it was a prelude to massacres of entire villages and the obliteration of whole societies.

So what do we do with this kind of ritual? How do we take the positive elements of harvest-time gratitude without glossing over the horrors of history? I am profoundly conflicted about it. But I do feel a very strong need to give thanks, and no qualms about finding a new way to do so.

It needn't be just once a year. In fact, it shouldn't be. Jon's grandparents are practicing Christians who see their faith as, among other things, a call to social and economic justice. At their table they say grace before a meal. When the family is all together it's often the singsong Selkirk Grace attributed to Robert Burns, the eighteenth-century Scottish poet, but sometimes it's the more traditional one. Both versions give thanks to God for putting food on our plates.

When I'm at their dinner table I can't help feeling that my gratitude lies elsewhere. Jon's granddad worked long hours for more than sixty years to earn the money that paid for this food. Jon's grandmother carefully, lovingly prepared it. And how many farmers, migrant workers, laborers, truckers, and supermarket employees made it possible for us to be eating this? If meat, poultry, fish, or dairy is being served, we've taken life or substance from an animal. I know Jon's grandparents feel this gratitude to everyone who helped make their meal possible, as well as to God. Still, I find myself wanting to give thanks to a long list of life-forms and biological processes out loud. And I'm not alone. Online there are dozens of secular ways to pray. Some are very derivative of Christian prayers. Some are creative and new. The Vietnamese Zen Master and peace activist Thich Nhat Hanh offers this:

In this plate of food,
I see clearly
the presence of the entire universe
supporting my existence.

William George Aston, an Irishman who spent much of the late nineteenth century studying Japanese culture, called Shintoism "a religion of love and gratitude rather than of fear, and the purpose of their religious rites was to praise and thank as much as to placate and mollify their divinities." What a difference that distinction must make on the everyday psyche of the believer. Among devout Shintoists, there is a ritual of closing one's eyes, bowing slightly, and, either silently or audibly, clapping in gratitude to the *kami*, the forces of nature, spirits, and other sacred phenomena that Aston called "their divinities." What a simple, elegant gesture a clap is to show gratitude. We use it so freely to show appreciation for a speech or a performance; why not just one for a meal?

In the last few years Jon and I have taken to a very informal version of this sort of thing. Mostly it's along the lines of "Thank you, farmers, thank you, fishies." This is in part for Helena's benefit. But feeling gratitude is not enough. Fasting is not enough. These rituals must compel direct action, as they do in, among others, the Mormon model.

In the online comments section of an essay I wrote, someone said, "When people are well-off, it's easy to accept a world in which this is all there is with no life after death." This is a very good point. Part of being secular must include an acceptance that any socioeconomic advantage you were born into was just random luck in the chaos. I would hope this would inspire a sense of moral duty to give to the less fortunate. No religion says *Don't feed the poor, God means for them to suffer.* In fact, quite the opposite. Many religious traditions strongly emphasize charity. This is arguably the very best thing about religion: a kind of social pressure to help others. But there is a secular argument for charitable works as well. If there's no rhyme or reason to why you grew up with three square meals a day, if there is not a great safety net of justice in the universe, we humans must create one for one another. Something like: *There but for the grace of chance go I.* Secular people must disabuse the rest of the world of the correlation between godlessness and immorality by becoming more focused on charitable giving and volunteering. The humanist hymns might take a while to catch on, but good works can start today.

In the times of hopelessness I've experienced about the state of the world—the inequity, the unfairness and oppression—in the moments when the world feels like a terrible place, I have found that there is also great solace in giving. Sometimes a monetary donation, sometimes a donation of time or effort. Marching for something you believe in, volunteering at an organization you feel is moving the world in the right direction, or boycotting a company that is not. These actions are more than wishes, they are work that has demonstrably changed minds and lives. And they solve another problem of secular life: they create a community, offering friendships with people who have similar values to yours, people to exchange ideas with, people to commiserate with, people to succeed and fail with, and people to fast and feast with. It's been crucial to our survival for eons. And no matter what comes next, we need a tribe to face it with us.

chapter fifteen
_____

Winter

There is no darkness but ignorance...
—WILLIAM SHAKESPEARE,
Twelfth Night

While the earth remaineth, seedtime and harvest, and cold and heat, and summer and winter, and day and night shall not cease. —GENESIS

The winter solstice, which falls at the end of December north of the equator and the end of June south of the equator, is the longest night of the year. Maybe more than any other day it seems to cry out to be marked. It is both our darkest hour and the turning point toward light, toward hope, toward spring again.

Once, on a late autumn trip to Japan as a kid, I was mystified upon seeing the center of the Ginza, the Times Square of Tokyo. We had been to several Shinto shrines (although I was excluded from some because of my young age, which I found very unfair) and the basics of the Shinto religion had been explained to me. Yet there in the Ginza the store windows were decked out for Christmas, complete with an enormous Santa, like one might find in New York or London.

"But I thought they're not Christian," I said.

"They're not," my mother told me. "They just like Christmas."

She was right. There are no two ways about it: Christmas is extremely popular. Even people who are neither devout nor even Christian love it. People often tell me it's just about being together, and doesn't need to be all wrapped up with Jesus. "It's literally called *Christ mass*," I say. I get eye rolls back. I'm being too literal, a stick-in-the-mud.

Several friends who immigrated here from non-Christian parts of the world—China, Iraq, Iran—have told me they didn't even realize it was a religious holiday at first. They saw it as an American holiday and they, as new Americans, felt as entitled to celebrate it as the archbishop of Canterbury.

I have two dear girlfriends who successfully lobbied their parents for Christmas. One grew up in a secular Jewish household and the other in a lapsed Muslim one. In one case, the parents felt conflicted about celebrating a holiday they didn't feel was theirs. They told the kids that if they hand-made a hundred unique paper snowflakes or similar DIY ornaments, they would get a tree, assuming the kids would give up after a few. But the task was accomplished with stunning speed and focus, and a Christmas tree was purchased.

People love the feeling of Christmas—the mood, the decor, the music—even if they have no connection with the belief system behind it. We all crave a little good cheer when the nights are long. I once heard a Satanist being interviewed on the news about his desire to have a Satanic symbol in his local town square next to the Nativity scene. The reporter asked him if he celebrates Christmas. He said, "I do, actually. I personally just view it as a time to be with my family." Even Satanists love Christmas.

And in the interest of full disclosure, I should say that when I was about fifteen I begged my mother for Christmas presents and a tree because I felt left out, and she acquiesced. Being a teenager, the urge to fit in, to be part of the majority, was sometimes overwhelming. Although I received a very flattering forest-green pullover that I wore for many years, I felt like a fraud. I didn't believe in virgin birth, I just wanted in on the fun. But luckily it turns out Christmas's roots long predate Jesus's arrival in the manger.

Even if we were to agree that every single word of the New Testament is literally true, nothing in the Bible suggests or supports the idea that Jesus Christ was born on December 25. If he was born on that day, it's an enormous coincidence. Biblical scholars have tried to deduce the actual date from clues in the sacred texts. Spring, summer, and fall birthdays have all been floated. Whenever it was, the commemoration of his birth wasn't established until about three centuries later, when Roman church leaders decided it should be on December 25. They did this so they could co-opt the existing polytheistic Roman holidays that centered around the winter solstice. (In the leap year–less Julian calendar, the shortest day of the year was on or within a day of December 25). This made things a little easier on Romans as they bid goodbye to their many gods and adjusted to life as Christians.

Previously, Emperor Aurelian, who ruled Rome for five years at the end of the third century CE, had been devoted to the sun god Sol. He must have felt that Sol was not getting enough credit, so he decreed that December 25 would be a holiday in Sol's honor called Dies Natalis Solis Invicti, or "Birthday of the Unconquerable Sun," which is an amazing pun about Christmas in English but, alas, not in Latin. It was a celebration of the changing of the seasons, the return of the light, the lengthening of days thinly disguised as the birthday of their sun god.

Before that, Saturnalia was a Roman gift-giving holiday honoring Saturn, the god of vaguely related things like abundance, capital, wealth, liberation, agriculture, cyclical renewal, and lead. An immigrant of unknown origin to the Roman Empire, remembered as Macrobius, wrote the best compendium of information on Saturnalia. Unfortunately he wrote it in the early fifth century CE, about a hundred years after Rome was Christianized. Saturnalia was no longer in vogue, and Christmas was not yet a full-blown holiday. During Macrobius's moment, he would have celebrated Dies Natalis Solis Invicti, but he studied Saturnalia as an artifact, a window into a lost world. So he wasn't writing from his own personal experience when he reported how Janus, the god of transitions—whose two faces look both backward and forward, and for whom January is named—"commanded that Saturn be revered with solemn grandeur of religious scruple, as the source of a better way of life." Macrobius tells of garlands and honey cakes and of overspending to impress people with gifts. Sounds a lot like today's Christmas, centuries before the Nativity.

Saturnalia had another element too, something closer to Halloween than Christmas. It was a kind of Roman "opposite day" on which the enslaved took on the roles of their enslavers, and men dressed in women's clothes, and their entire rigid power structure was, for one brief moment, turned on its head. After all, the solstice itself is also a kind of opposite day, the moment when the paradox of extremes in northern and southern hemispheres—long summer days contrasted with short winter ones—is most extreme. It's almost irresistible to draw some parallel to the beauty and tragedy of our short, small moment on Earth.

According to my old friend the *Encyclopædia Britannica*: "The use of evergreen trees, wreaths, and garlands to symbolize eternal life was a custom of the ancient Egyptians, Chinese, and Hebrews. Tree worship was common among the pagan Europeans and survived their conversion to Christianity." Humans have been decorating trees, particularly evergreens, for eons, for good luck and as reminders of what spring will bring. That's why ornaments look so much like fruit.

Another precursor to Christmas was the birth of Mithra, the god of light of Persian extraction. Like Christians, the followers of Mithra believed their god was conceived without sex and born in late December. (Although this conception involved coming out of a rock.) Other virgin birth stories have been part of belief systems on every continent, from the Aztec earth mother goddess Coatlicue (who conceived her son Huitzilopochtli, a war god, when a ball of feathers fell from heaven) to the Yellow Emperor of China (whose mother supposedly got pregnant when she was exposed to a lightning bolt from the Big Dipper in the twenty-seventh century BCE).

The Druids used mistletoe in their ceremonies and connected it to, among other things, virility and fertility. It's unclear if the modern custom of kissing beneath the plant at Christmas parties is a PG-ification of that, but there appears to be a theme here.

None of this makes Christmas any less legitimate. More than any other time of year, we need a holiday in the dead of winter. We need something to remind us that darkness is not forever.

Hanukkah, no surprise, is rooted in the same struggle between darkness and light. Since the Han dynasty the people of China have marked the solstice with the consumption of little rice ball dumplings and the honoring of ancestors, called Dongzhi. The ancient people of the Punjab region of India marked it with an enormous bonfire on Lohri (which is still celebrated today, but over the centuries has moved from the end of December to the middle of January and now features children who sing door-to-door for little treats). There is Yaldā, which is celebrated in Iran, Azerbaijan, Afghanistan, and most of the other central Asian nations that end in the suffix -*stan*. Beloved poems, particularly those of Hafiz, are read aloud by candlelight. This pre-Islamic, Zoroastrian holiday originally included the ritual of staying up all night to keep evil spirits away. Today it's mostly secular, but people still stay up late. Someday other holidays in late December might be invented, and our current ones will be relegated to the history books or altogether forgotten.

When Jon and I got together, we each came into the relationship with our own late-December gift-giving traditions. Since he grew up with Christmas and I with Hanukkah, part of our coming together, our becoming really serious about each other, was deciding how to navigate this. What did we really want to celebrate? What did we actually care about? The second December of our cohabitation, we were living in London and decided not to go home to see our families. Instead we took the Chunnel to Paris and spent the holidays snuggled up in a small rented apartment in Montmartre. It was there, away from all the familial holiday pressure, that we first felt free enough to make our own traditions. Based not on what we were used to but on what we actually believed. Without a tree or menorah, we drank cocktails and exchanged gifts at midnight on the winter solstice.

Long before Helena, before we were married or even engaged, when we still danced around the idea of spending the rest of our lives together, that trip to Paris was when we first talked about the traditions we'd make for some future hypothetical child. Maybe at the stroke of midnight on the longest night of the year, we imagined, we'd slip into their room, gently wake them up by whispering into their little ears, *We have something wonderful, something epic and thrilling to tell you, something so large and magnificent that no human being can stop it: starting tomorrow the days will start getting longer again, and slowly the plants will bloom again, the sunshine will return. Summer is coming.*

Then, with the aid of a diagram or a globe and a flashlight, we could teach them why this is so, how this miracle actually works. We'd ask them to imagine the people in the southern hemisphere who are, at this very moment, enjoying the shortest night of the year in the midst of their summer. Tomorrow their days will begin to shrink, their winter will slowly approach, and in six months we'll have traded places. We dreamed of telling them it's a system almost 5 billion years in the making.

And then, a feast and presents. Because what is more exciting when you're a kid than getting to be up late at night with your parents?

Helena is too little to feel one way or another about this idea. She and any future siblings may be greatly disappointed by being the only kids in the neighborhood not, for example, being visited by the suggestion of a large man who creeps into their house at night via the flue. That doesn't mean, however, that we can't honor the ancestors who were believers. Helena has a lot of them. She contains the genetic material of people who came to America because of their beliefs. Jon is a direct descendant, on his mother's side, of John Robinson, the English Separatist known as "the pastor of the Pilgrim fathers," who cofounded the Congregational Church. Robinson's flock awaited him in the New World, but he never made it. He died in exile in the Netherlands, never able to cross the ocean himself, to see the land that would one day become a nation, in part, because of the theology he preached. But his son, Isaac, arrived in Boston in search of religious freedom. And because of that, and a virtually infinite number of other decisions and chance meetings, Helena is here now. On Jon's father's side, there are family stories of Mennonites who also fled persecution in the Old World. My ancestors, of course, almost all came in steerage, through Ellis Island, desperate to escape the shtetls and pogroms of Eastern Europe.

During the first December of Helena's life, she got a lot of presents, mostly board books. She got one each night of Hanukkah, after we lit our ultramodern, almost abstract menorah despite two impeding factors: some kind of Hanukkah candle shortage in our neighborhood that forced us to use fancy birthday candles not meant to burn for hours, and our complete inability to retain the very short Hebrew prayer, needing to look up the phonetic version every night. Repeating declarations of worship that I don't agree with is so much easier to do when they're in a language I don't understand. Despite my own nonbelief, some part of me had the sense that I ought to give this experience to Helena, not as a directive but as an option. But this was not only for her sake. I cannot deny feeling a vague, ancient, wholesome warmth when I follow in the footsteps of my ancestors. Some deep part of me doesn't want to be the one to break the chain. Eventually Helena will have to sort out how she feels about all this, but for the time being she seemed to like watching the little flames.

On the winter solstice that year we found ourselves in Ithaca, at my mother's house. It was not the plan we had envisioned, but as I nursed Helena in the wee hours I explained the movements of the planet, and she looked up at me, practicing expressions of understanding. Or maybe she really did understand the most basic idea: Something good was coming. Things would be getting better.

On Christmas Eve we went to Jon's mom's house. Jon's dad, his wife, and their daughter came over. I have spent many Christmases with Jon's family, complete with ornamented trees and personalized stockings. It's fun. But I can't help feeling I am the odd woman out. Not because their Christmas is the least bit theistic. It is as secular as anything we celebrated growing up. But because Christmas just doesn't come naturally to me. I still feel I am learning it. It reminds me of my otherness. It makes me feel like I'm pretending to be Christian, trying to pass as a member of the majority when I know deep down I am not. It's not so different from how Jon felt at that first Seder.

On that first Christmas Eve of Helena's life we were discussing our plans for leaving the next morning. My mother-in-law suggested we do the ritual present-opening that night, the evening of the twenty-fourth, so we might get on the road a little sooner, to beat a snowstorm and make it to Jon's grandparents outside of Boston in time for Christmas dinner. A completely reasonable plan. But to my astonishment, a little part of me felt sad. Not for myself but for Helena. She is, after all, as entitled to Christmas as she is Hanukkah and whatever winter solstice celebrations we've come up with. She has as much Christian ancestry as Jewish. I found myself thinking, *She has to have Christmas morning!*

"How about if we do just Helena's presents in the morning?" I asked my mother-in-law.

She agreed and we did. I even put her in some red-and-green pajamas friends had given us. "Who am I right now?" I asked Jon. Maybe just a mom who wants her child to have everything wonderful there is under the sun.

Withholding Christmas from her would have been a kind of censorship. I would be doing her a disservice by only emphasizing the holidays that I felt a connection to. Only through experience can she decide how she feels about them. If my parents had no qualms with me, who has no Christian ancestry, attending church with Maruja, how could I deprive Helena of Christmas?

After she opened her presents (with a lot of help), we put her upstairs in her travel crib while we packed up the car for the almost-six-hour drive ahead. From the kitchen I heard my mother-in-law yell. Jon and I raced up the stairs to find Helena on her belly, head and shoulders raised up like the

Sphinx, smiling proudly. She had rolled over for the very first time. A little Christmas present from her to us.

I will never tell Helena that a big jolly man is coming down the chimney to eat cookies and leave her gifts. I won't tell her that her grandpa Carl is watching over her. Or that there is a man who created everything and lives in the sky. But I will also never stop her from choosing her own path. I don't know who she will be or what she will believe. But it will be up to her. Authority—a parent or a church or a government—cannot enforce belief. Or lack thereof. She will have to be true to herself. Because "the only sin would be to pretend."

On the scale of our human history, rituals like putting up Christmas trees, lighting menorahs, reading Hafiz, and baking rice dumplings are new. We humans have celebrated the earthly repercussions of our orbit longer than we've celebrated virtually anything. Before Christmas and Hanukkah, before monotheism or any other kind of theism, our ancestors were staring up at the stars, trying to gather clues about the changing of the seasons, the passing of time, and what the darkness might bring. The idea of marking the longest, coldest night with the knowledge that the warmth and light is not too far off, *that* is ancient. And no matter where we're from, what religion we are, or to what ethnic group we belong, we can be sure that our ancestors, *all* of our ancestors, contemplated Earth's place in the universe with awe.

For them, it was sacred. And it still can be for us. Even more so because science has brought us a deeper understanding of the mystery and beauty of nature than our ancestors could have ever dreamed.

Usually the words *mythology* and *myth* imply inaccuracy, but as Karen Armstrong writes in her book *A Short History of Myth*, "A myth was an event which, in some sense, had happened once, but which also happened all the time. Because of our strictly chronological view of history, we have no word for such an occurrence, but mythology is an art form that points beyond history to what is timeless in human existence, helping us to get beyond the chaotic flux of random events, and glimpse the core of reality." In another of her books, *Fields of Blood: Religion and the History of Violence*, she wrote, "A myth is never simply the story of a historical event. Rather it expresses a timeless truth underlying a people's daily existence. A myth is always about now." Every holiday we celebrate, every birthday and independence day, everything is as much about the present as it is about the past. We are both constantly changing and constantly repeating our oldest patterns.

Armstrong calls this concept "everywhen." Nothing could be more "everywhen" than the revolving of the Earth around the sun. It's the closest thing to a literal everywhen we are likely to experience. It's been going on for 5 billion years, and in all likelihood will continue for billions more.

I've spent a lot of time talking about how great the winter solstice is, how worthy it is of celebration. I've bored people at dinner parties about it. But there's something I always omit. The winter solstice is also the anniversary of my father's death. He died in the early hours of December 20, when the stars shone the longest.

*

chapter sixteen
───────────────

Death

Don't gloss over death to me in order to console me.
—HOMER, The Odyssey

One day, of course, no one will remember what I remember.
—DONALD HALL

On the first December 20 of Helena's life, the twenty-first anniversary of her grandfather Carl's death, we drove with her to Ithaca for the first time. Her "ancestral home," we jokingly call it. The place where Jon and I grew up, met, and, years later, married. It's where her three living grandparents reside, and where the fourth, my dad, is buried. It is the home—or at least hometown—of many of Helena's biological and honorary aunts and uncles. And we're almost positive it's the place where she was conceived.

The next day, December 21, was the winter solstice. That morning we took Helena to the cemetery for the first time.

Among the Jewish traditions I observe devoutly in my own secular way is the placing of a small stone on the grave of a loved one. When I go to the cemetery where my dad and grandparents are buried, I have, for many years, picked up five small stones. One for my dad, one for my grandma Pearl, one for my grandmother Rachel, one for my grandpa Sam, and one for a boy I knew and loved in elementary school who was hit by a bus while riding his bicycle in 1991. As I think about it now, he might have been the first person I really knew, really still remember, who died. Certainly the first who was not old. Now I collect six small stones when I arrive at the cemetery. One more for Grandpa Harry. Walking down the hill toward Helena's ancestors, we look out over Ithaca, not quite able to glimpse the middle school where Jon and I first met, Cayuga Lake, and much of our hometown, built on somebody else's sacred burial grounds. Some other girl's father's grave.

My family is not buried in the consecrated Jewish section of the cemetery, as my father's parents would have probably preferred, but just over a little chain-link gate, as if eternally waiting beyond the velvet rope, never to actually enter the club. Which is probably the most poetically perfect place for us. Not exactly Jewish, but Jewish adjacent. I say "us" because I too plan to be buried there. I have this sense that I would be "happy" there, and that I would be "unhappy" somewhere else. I have no reason to believe it will make the slightest bit of difference to me once I'm dead. But maybe it will give me some comfort just before, should I meet the kind of end that gives some warning. Maybe I'll think for that last split second, *At least I'll be in that pretty spot overlooking Cayuga Lake, near my dad.* Although once I'm there, will it really be *me*? And is it really my dad that I'll be near? Or just the remains of the casing we once wore? It's very hard to grasp from this obscured vantage point of being alive.

At my father's grave there are more than pebbles left to acknowledge his absence. People from all over the world leave notes, marbles, Lego, mini planets and other space-related objects. Seeing this makes me happy. My grief is soothed by knowing other people miss him, remember him, and still love him now.

Standing at his gravesite so many years later, the clearest memory of his funeral is the feeling of staring at his coffin, unable to fathom that my dad was *in there*. It's not sadness I feel when I think about it but astonishment. I cannot wrap my mind around it. All these people making their way around the planet, the people you love, the people you hate, the people you know, the people you hardly glance at, fully animated, and then suddenly—or slowly—the animation goes out of them. Or rather, out of us. And only the body is left. And then some of us put that body in a wooden box and put it in the ground.

I had taken Jon to "meet" my family there many times while we were dating. He seemed very cool about it and not as weirded out as I had feared. The first time Jon and I visited the cemetery after we got engaged I realized that I too had a proposal for him. "Now that we're getting married, will you be buried with me here?" He accepted. I was overjoyed. Everything seemed bright and beautiful. I might be conflating visits, but I think the lilac bush next to my grandpa Sam was blooming. It all seemed so romantic, like the deepest possible commitment, this pledge to stay together even after death, that I blurted out, "I can't wait!" Which, naturally, sort of horrified Jon.

I, of course, *can* wait and would like very much to wait as long as possible. Until then, I get some pleasure from the idea that this is where we'll be. Not for eternity, because this planet will be absorbed by the death throes of the sun in about 5 billion years. And, in the nearer term, the enormous amount of land used to house dead bodies in countries like the United States will likely be needed for something else as soon as people stop thinking of those of us buried there as their grandparents and starting thinking of us as somebody else's. (Like how the Europeans viewed sacred Native American burial grounds, for example.) But nevertheless, I derive pleasure from the idea of being buried there with my loved ones. The idea of it is beautiful to me. I also derive pleasure from freaking Jon out a little now and again. After the day he agreed to be buried in our family plot, I started joking that that wasn't quite close enough, that what I'd really like is for us to be buried together in the same coffin. It might have started in part because of a plot shortage due to an embezzlement scandal at the cemetery, resulting in some plots (not ours thankfully) being sold to multiple families. But it soon seemed like a genuinely romantic idea. A few years later, it became clear; as with any idea, I was not the first person in history to think of it. One day at the Museum of Fine Arts in Boston, we happened upon the coffin of the late Mrs. Thanchvil Tarnai and her husband, Larth Tetnies. They loved each other during the fourth century BCE in present-day Vulci, Italy, a town about equidistant between Florence and Rome. And their enormous sarcophagus is sealed by a lid featuring larger-than-life versions of themselves, seemingly naked, entwined in a sheet, holding each other, staring into each other's eyes, the very picture of two people waking up in an embrace after a night of passion. The title of their museum card reads, "Together Forever."

I imagined them in there, still together, bones intermingling, closer even than during sex, a possibility only after it can no longer be enjoyed. But after many a visit, a curator disabused me of this image with the words *No remains were saved*. Thanchvil and Larth are "together forever" only in the image we have of them, here, across a vast ocean, in this impossible future.

There is a school of thought in anthropology that argues that our humanity began when we started ritualizing death, rather than walking away and letting one another decay out in the open. Putting the bodies of our loved ones in boxes under the earth with a small stone sign that gives their name, the dates of their birth and death, and maybe one small detail about them, happens to be our custom, but it's no more or less strange or rational than any other. Cremation is both ancient and increasingly popular. Nowadays we do the cremating mostly behind closed doors, a ritual hidden from the grieving. But public funeral pyres, where you watch the body of your loved one become engulfed, have existed for millennia, and still exist today. From Buddhist monks in Laos to hippies in Colorado, humans sometimes say goodbye with this kind of bonfire that bonds you to your loss.

Well beyond ancient Egypt, around the world and throughout history, humans have mummified their loved ones, a way of preserving their bodies a little longer. In Tana Toraja, a remote, mountainous regency of Indonesia, locals today still grieve and honor their dead by keeping their bodies out in the open, mummified, but often treated as though they are just sick, just quiet, not dead. Something not so different from the dioramas at the American Museum of Natural History. Maybe it's the perfect physical incarnation of the first stage of grief: denial.

Among scattered communities from Vietnam to Madagascar, and for the indigenous people of the Amazon and Great Lakes, burial is not just a onetime thing. Years after the shock of loss, once a body has decomposed, it is exhumed. Some customs call for the bones to be lovingly cleaned and reburied. Maybe this is akin to some kind of closure.

I think I would have greatly benefited from living in a society where all the stages of grief were spelled out in ceremony. Not that I would have liked to sit with my dad's mummified corpse in the dining room as I finished high school. But I craved some ritual process by which my community, my classmates, my teachers, and my family could more openly acknowledge how long and difficult this was going to be.

At my dad's funeral, a family friend—I'm sure believing she was being helpful—gave me an antianxiety pill of some kind in case I "got too sad." I took it about halfway through the eulogies. I regret this now, more than two decades later. I'm certainly not against antianxiety medicine, but I was too young to know what I was doing. I needed to feel the terrible feelings of that day. I didn't realize that I needed to feel sad, even "too sad," that this was a crucial opportunity for group catharsis.

A friend told me that when her aunt died the imam performed a Persian tradition where he theatrically, dramatically willed himself into a fit of tears, cajoling the mourners to do the same, to let out their feelings, to feel comfortable weeping for their dead. I needed something like that.

Secular though we are, we read the mourner's kaddish, the Jewish funeral prayer, at my dad's grave when we buried him, and twenty years later when we buried Grandpa Harry, just feet away. We did it because we don't yet have something else to read that both represents our values and lends the gravity of tradition. Maybe by the time I go I can leave my loved ones some list of passages that might better reflect my beliefs. But there is at least one Jewish funerary ritual I would like included. It is one that involves no words. When the coffin is lowered into the ground, it is the family, the loved ones, the community, who take turns, one by one, picking up the shovel and casting down the dirt that buries their dead. As with so much, Jews and Muslims are more alike than different. In Islam it is three handfuls of soil.

"It's the sound that kills me," Jon said as we left my great-aunt Frieda's funeral years ago. A series of thumps that signal finality. Of all the turning points in a life, death begs most for ritual. Many of these are elaborate, but the simplest ones sometimes carry the most weight.

For affluent ancient Egyptian women of the Late Period (about 664 to 332 BCE) who had lost a male relative, the mourning process included walking through the streets beating their bare chests. In parts of modern Bulgaria, when someone dies you put up a poster of them in town. It showcases their picture and a few details about their life. The poster goes up again forty days later, and again every year, so people remember the person you've lost, so you don't have to grieve alone. Jews cover the mirrors in their homes with fabric during the first few days of mourning. The idea is you shouldn't have to worry about how you look when you're grief-stricken. For most of my life, I knew the custom of pouring out a drink for a dead friend who should be there drinking with you only from the rappers I grew up listening to, but it's ancient. In ancient Africa, Rome, and Greece, among the Inca and the Chinese, there is a history of pouring one out for your dead, for your ancestors, or for your gods.

Upon first reading, hearing, or seeing each of these rituals, something clicked inside me. *Yes*, I thought. *These traditions require no belief.* To me, they seem so clear and direct I almost can't believe everyone doesn't do them. Maybe these specific rituals don't speak to you, but for each of us there is, I hope, some ritual that helps interpret this great mystery we all must face.

We pretend these rituals are for our dead, but they are for the living, for us.

It was cold and gray and the trees were bare when we brought Helena to see the family headstones at Lake View Cemetery, but I was reminded of the sunny morning of my wedding, when I had snuck off to weep alone there. I didn't cry this time, but Helena did. Maybe she was just hungry or tired—she was less than five months old at the time—but it was impossible not to ascribe my own deep emotions to her little cries.

Every loss you withstand in your life reopens all the others. Every goodbye is every goodbye. My granddad's death reopened the death of my grandmother eleven years earlier, and the death of Maruja, who I hadn't said a proper goodbye to, even over the phone, which eats away at me still. My friend Brent, who died in 2001 and whose voicemail greeting we listened to again and again, who I was too scared to say goodbye to in his hospital bed, even though all our other friends did. The mother of another lifelong best friend, who had died just a year earlier. Our whole group of girlfriends came home to Ithaca and got into bed with her to say goodbye. All these losses feel connected. They all pull me back to the original heartbreak of my life: the loss of my father.

One of the last things he said to me was "I'm sorry." I could not understand, for many years, why he could possibly be apologizing to me. I should have been apologizing to him. He was the one who was in so much pain. He was the one who was dying. But he understood what I was too shocked to grasp. This would be the defining event of my life. Every other loss, every other heartbreak, would reopen this wound. And even the very best moments of my life—any future successes, my wedding, holding my newborn baby for the first time—would be tarnished by his absence.

I often think back to that first conversation I had with my dad about his parents, why I had never met them, and the existence of death. I think about my attempts to elicit promises of immortality from him before bed and being met only with "I'll do my best." Because he had been sick off and on for two years, when he really was dying I was not aware that he might not make it until the last few days of his life. My mother, who was and is a devout optimist, focused on the good news and not the bad. We believed he would be okay. In a way, the "I'll do my best" mantra was the best preparation I'd had. And he lived up to his promise. He did do his best, but it wasn't enough.

My parents taught me that even though it's not forever—*because* it's not forever—being alive is a profoundly beautiful thing for which each of us should feel deeply grateful. If we lived forever it would not be so amazing. It doesn't mean loss isn't scary, it doesn't mean it's not hard. But for me, it has helped.

In the fourth century BCE, the Chinese Taoist philosopher Zhuangzi had to face this when his wife died. His friend came to pay his respects and found him singing, drumming, and flouting the social norms of grief. The friend couldn't understand why Zhuangzi wasn't sad. Zhuangzi told him that his wife's life had evolved into her death "like the four seasons in the way that spring, summer, autumn and winter follow each other." He said he thought about the time before his wife was born and what a miraculous wonder it was that she had been born at all. Death was natural to him, part of the system of *tao,* or "the way," part of the working of the universe. But he also told his friend that he had mourned, that he had felt despair, too. How sad he still must have been, despite his belief that this was the way things must be.

Just after Helena was born, as I listened to the audio-book of my dad's *The Demon-Haunted World,* I came to his account of his goodbye to his own dying father, my grand-father Sam, for whom my brother is named, but who we never knew. He found himself saying "Take care," as though his father would go on to some other life where that would be helpful advice.

For me, the original Sam Sagan is a composite of a few black-and-white photos, a list of positive adjectives, a hand-ful of anecdotes, and the headstone beside my father's. He is not a flesh-and-blood person. But he was to my father, and as I stood at my father's grave with Helena, I realized that my dad will never be a real person to her. But he was real. And so was my grandfather Sam. And when they were small they were told stories of dead relatives that they would never meet by people who knew them, loved them, missed them. And on back beyond the horizon of time to the be-ginning of language. Or at least to the beginning of gram-mar allowing for past tense. Before that there was probably

some incommunicable longing, some early form of mourning, unsaid but palpable between creatures not yet human.

Our long-lost cousins the Neanderthals buried their
dead with the odds and ends they felt were important. Important why or how is, of course, lost to time. These people-
ish humans who last walked the Earth some forty thousand
years ago sometimes buried their mothers, fathers, sisters,
and brothers in the fetal position. Maybe to prepare them
for some other life. Maybe just to acknowledge in some prehistoric way the circularity of everything. The Neanderthals were innovative, after all. They seem to be the first
creatures on our planet to do anything with their dead.
What does that say about them? Or us?

In *Breaking the Spell*, Daniel C. Dennett wrote, "Pulled
by longing and pushed back by disgust, we are in turmoil
when we confront the corpse of a loved one. Small wonder
that this crisis should play so central a role in the birth of
religions everywhere."

Many cultures have a structure in place to help people
face death, and if not that, at least to know what to say
when someone dies. For those of us who do not believe,
death can be very awkward. After my dad died people said
the strangest—sometimes very upsetting—things. Several
people told Sam he was "the man of the house now." He was
five. Imagine the stress that might cause a child who can

barely wrap their mind around what has happened. Imagine the insult to my mother, now a single parent to two children, being told her kindergartner is in charge.

Even decades later I have awkward exchanges. A few years ago, when I was well into adulthood, someone asked me how long ago my father had died and I told him it had happened in 1996. He replied, "Oh, so you're pretty much over it now." I burst into tears. I've more than once been in a situation where someone recognizes my surname, asks me if I am related to my dad, and describes their admiration for him in the present tense. Sometimes I go along, not wanting to break the news of his death to a person who lives in the world I wish I could be part of, where my dad is not dead but just elderly and quietly teaching classes, maybe, no longer interested in being on TV. But sometimes a question is asked or a request is made, like "Please tell him how much I love his books," and I have to deliver the news of his death, most recently, twenty years late, to a man selling me eyeglasses. I hate to be the bearer of bad news, but what makes it so hard is how awful the other person feels. Through absolutely no fault of their own they've brought up the most painful thing in my life and they are often mortified. I feel terrible for them. It's the kind of thing I would do too and then cringe about it while trying to fall asleep for the next ten years. So I do my best to relieve the awkwardness, saying things like "It's okay," which is a lie. It is not okay. It's very sad, but we don't have a way of talking about this without making it weird for everyone.

In the months after my dad died, someone we knew brought her toddler by the house. After the little boy left, my mom turned to me and said, "I just realized it's like you die again when the last person who knew you dies." The boy was probably the youngest person my dad knew at the time of his death, younger even than my little brother, who was just shy of six.

I didn't know this then, but the Ovambo of Namibia have words for this concept. It is, as Professor Olupona wrote, "a distinction [that] exists between 'ancestors' and the 'living dead.' The former (*aathithi*) refers to the 'forgotten deceased,' those whose activities and memory cannot be recalled by the living members of their lineages. The living dead, on the other hand, are the recently deceased."

I don't know if Maruja ever wanted to have children of her own. She never married. I don't think she was ever in a relationship, but I don't know for sure. And yet I can't help feeling that I, like the other children she helped raise, am a continuation of her in some way. We don't have her DNA, but we carry what she taught us, how she influenced us, and we can pass that down like language, like culture, like ritual. This is true for everything handed down to us by people we are not related to.

Memento mori is Latin. It means "Remember you have to die." It was used by the Christian Church of the Middle Ages to remind parishioners that the temptations to sin during your stay on Earth will, in time, catch up with you. You will meet your maker, and the lust, gluttony, greed, sloth, wrath, envy, and pride you got up to here on Earth will not seem worth it. But for those of us who do not think there is a piper to be paid, or at least have different definitions of what makes right and wrong, we must still *memento mori*.

You and I and everyone we know will die someday. Yes, our species will end or evolve into something unrecognizable to us. Yes, the sun will die. Life on Earth will come to a close and many more things we have not yet predicted will take place in the universe. And if the universe keeps expanding like it is, in about a quadrillion years the last star will have died. And that's the best-case scenario. It's very easy to let that get you down. It's scary. It's depressing. I'm anxious just typing it. But it's real.

I hold open the possibility that there is something on the other side of death besides dreamless sleep. There is so much we don't yet understand. So much our species will likely never understand. But in the short time we do have, there is so much shocking beauty and profundity that we have managed to grasp. We must find a way to relish the inner workings of evolution, the human brain, the universe, photosynthesis, biology, reproduction, gravity, genetics, and physics as we see them, to celebrate how dazzling nature really is. If there turns out to be something more than dreamless sleep after death, the way it actually *works*, the why and how, will be part of nature, not something "supernatural." So much of recent human history has been a process of taking individual phenomena out of the magic or religious column and putting it in the scientific column: disease—especially mental illness—droughts, floods, earthquakes, weather, the layout of the solar system, the abundance of flora and fauna. Somewhere in that process we lost the awe. I suspect this was partly a matter of delivery. We don't teach children science (or math, for that matter) with the passionate enthusiasm of the best preachers. And we ought to.

There are a lot of T-shirts, posters, and memes in this world that say, *Somewhere, something incredible is waiting to be known.—Carl Sagan.* He did not say this. It came from a *Newsweek* profile on him and somehow took on a life of its own. My mother and I find this slightly funny and slightly frustrating. My dad was so committed to accuracy that he would not have used the word *incredible*, because it literally means "not credible" and therefore the opposite of the kind of thing that's out there waiting to be known. There are, however, surely credible things out there, somewhere, waiting to be known. Credible, stunning, splendid, bewildering things. Some maybe just around the corner. Others waiting for Helena's children's children's children to behold. More astonishing things will be revealed in time, not just by our individual experiences but by the scientific method, by a deeper understanding of the mechanisms at work in our universe, by scrutinizing and testing concepts until they can become theories. Whatever it is that we have yet to learn will be part of nature once we understand it. And when we do, I hope we can still feel wonder. In those revelations and the ways the randomness, the chance, the chaos sometimes, somehow works out. Still magical. Still beautiful.

No matter what the universe has in store, it cannot take away from the fact that you were born. You'll have some joy and some pain, and all the other experiences that make up what it's like to be a tiny part of a grand cosmos. No matter what happens next, you were here. And even when any record of our individual lives is lost to the ages, that won't detract from the fact that we *were*. We lived. We were part of the enormity. All the great and terrible parts of being alive, the shocking sublime beauty and heartbreak, the monotony, the interior thoughts, the shared pain and pleasure. It really happened. All of it. On this little world that orbits a yellow star out in the great vastness. And that alone is cause for celebration.

A Postscript

One day when Helena was almost seven months old—teething, learning to eat real food, getting the hang of peek-a-boo and coming into her own identity—my mom called. I remember I was holding Helena and it was hard to focus on what my mom was saying.

"Would you like to play your grandmother Rachel?"

"What?"

"I just got out of a meeting. And there's going to be a scene in the next season of *Cosmos* of Dad as a little boy, drawing his dream map flier."

As a child in his Brooklyn apartment my dad had drawn an imaginary flier from the future, inviting civilians to sign up to join an interstellar fleet. It's adorable and a little prophetic. His careful block lettering, the narrow design of the rocket, the tiny face of an astronaut all fill me with love for a little boy I only knew as a middle-aged man. I love it so much that I've used a photo of this drawing as the lock screen on my phone for almost five years, since around the time that the flier and many other things my parents created were acquired by the Library of Congress the week we were visited by the singing cab driver (and possible deity).

"You mean the lock screen on my phone?"

"Exactly."

She reassured me it wouldn't be a speaking role and would only take a day. Before I could even process what this would entail—flying across the country with Helena, being on camera—I realized I was telling my mother I would do it.

I know TV is pretend. I have even been on a few sets over the years. And yet it seemed like being on a soundstage, putting on a costume, and pretending to be my grandmother, watching over a little boy who was pretending to be my dad, was going to evoke something real in me. I wanted to be Rachel for a day, to time travel to the Brooklyn apartment, see my dad as a little boy. I wanted to witness him not as a dying genius but as a child full of wonder, with a Brooklyn accent and his whole life ahead of him. It wouldn't be my dad, of course, but a kid I'd never met before cast to play him.

The day before we left, I put on the only thing I have of Rachel's. Well, not the only thing, but maybe the only object. Not counting laughs, not counting features, not counting a quarter of the information that flows through my blood. Beyond those heirlooms I only have one—a ring. When Rachel married my grandpa Sam, they used a simple metal band, which my cousin Sharon, Rachel's only other granddaughter, wears as her wedding ring. Many years into marriage, Sam bought her a new ring, a band of rubies, which I was eventually given by my mother. I put it on my finger and tried to time travel. Not to her unwrapping it, because that would happen years later, but to her wanting it.

But the ring wasn't enough. I needed more Rachel if I was going to take this seriously. I remembered that there was a reel of silent home movies, scenes strung together over a decade or two on the Library of Congress website. I watched it on my phone the night before we left, after Jon and Helena were asleep. I watched Rachel. I did look a lot like her. Not as much as my brother looks like my young dad, sweet and funny. In bed I tried to copy her hand gestures because that seemed like something actors do. I watched how she watched my dad, admiringly, lovingly, slightly amazed. Like I watch Helena.

Getting on the plane the next day, I wondered how old she was the first time she flew.

At the studio, in the buzz of production, I remembered

that, besides being the woman who gave me life, besides being a producer, writer, and director, the creator of a huge network series, my mother is still a wife carrying on the legacy of her true love.

"This is where Grandma works," I told Helena. "And Uncle Sam." My brother is a co-producer.

As I had my makeup done, as I was being transformed into my dead grandparent like the grandsons of the Chinese Axial Age, I read the scene. At the top of the scene it said, "We are in a memory," words that, of course, my mother wrote. Not just any memory, not an anonymous memory, like so much of history, but my dad's memory. Not perfect, not photographic, not even firsthand. An approximation of something that once existed and was then stored on the neurons in his brain that now decomposes under the earth. But it's a memory that was described to another person, my mother, who lives, and has built a world where it can survive a little longer. It was real once, there was such an apartment in 1946, where my dad, his parents, and his sister lived. And I was about to get as close to it as I humanly could.

A crew member led me onto the soundstage, a vast room with high ceilings. I could see SOTI, the Ship of the Imagination, in the distance. My mom took me into a small doorway. As we passed through we both touched the mezuzah, the traditional Jewish prayer parchment in a special case, hung on the doorframe as part of the set design. It's something practicing Jews always do, but not us, usually.

And with that we were through the portal, into my father's childhood apartment. I was trying to be professional, but I was overcome with emotion, fighting back tears. It was almost time travel. Among the doilies, mail from the War Department, the spice rack and menorah, there was a photo of an elderly couple, two Orthodox Jews from Eastern Europe. My great-great-grandparents, who starved to death a hundred years ago, were there, too.

The healthy young man playing my adolescent dad was lying on the floor. From my vantage point in the kitchen I could pretend he really was my dad and I really was Rachel. When the director called, "Action!" I performed her small daily tasks, trying to imagine her emotions, her secret inner life, the way she saw the world. Imagining my father's future, our past.

Jon brought Helena by to see what we were doing between shots. My brother poked his head in to watch, too. And there we all were, sharing different versions of the same names—Rachels, Sams, and Chayas—each a part of one another, bending time and space to be together, like something out of Gabriel García Márquez's *A Hundred Years of Solitude*, where many generations of a family exist simultaneously, living outside the boundaries of time.

A few weeks later I noticed Helena staring at me from

her highchair while I steamed some vegetables for her—food grown in the earth, with the help of our nearest star, ready for Helena to convert into energy to grow bigger as a result of millions of years of evolution. Quizzical, observing, she watched me do my small, daily rituals. These images of me, performing semisacred tasks, were being imprinted into her brain, into her memories. Someday I will be gone, but she will, I hope, remember me. And I'll be able to live on, just a little longer, in the neurons in her brain and the cells in her blood.

Suggested Reading

Every day, the number of books I have not read gnaws at me. There are so many brilliant ideas out there and I have so much still to learn. If you have recommendations for me, please find me on the Internet. I would love to hear them. In return, here are some nonfiction works I have read and loved or been enlightened by or feel fit into the themes of this book in one way or another.

1491: New Revelations of the Americas Before Columbus by Charles C. Mann

Battling the Gods: Atheism in the Ancient World by Tim Whitmarsh

Between the World and Me by Ta-Nehisi Coates

SUGGESTED READING

A Brief History of Time: From the Big Bang to Black Holes by
    Stephen Hawking
Dear Ijeawele, or A Feminist Manifesto in Fifteen Suggestions
    by Chimamanda Ngozi Adichie
Fields of Blood: Religion and the History of Violence by
    Karen Armstrong
*From Here to Eternity: Traveling the World to Find the Good
Death* by Caitlin Doughty
The Great Courses series
*The Great Transformation: The Beginning of Our Religious
Traditions* by Karen Armstrong
*The Invention of Science: A New History of the Scientific
Revolution* by David Wootton
Marriage, a History: How Love Conquered Marriage by
    Stephanie Coontz
*A Million Years in a Day: A Curious History of Everyday Life
from the Stone Age to the Phone Age* by Greg Jenner
*Our Magnificent Bastard Tongue: The Untold History of
English* by John McWhorter
*Religion for Atheists: A Non-believer's Guide to the Uses of
Religion* by Alain de Botton
The Religions of the American Indians by Åke Hultkrantz
The Rites of Passage by Arnold van Gennep
Seven Brief Lessons on Physics by Carlo Rovelli
*She Has Her Mother's Laugh: The Powers, Perversions, and
Potential of Heredity* by Carl Zimmer

A Short History of Myth by Karen Armstrong
Time Travel: A History by James Gleick
Very Short Introductions series

And, of course, my parents' work is the source of endless inspiration for me. Some places to start might be:

Cosmos
Pale Blue Dot: A Vision of the Human Future in Space
Shadows of Forgotten Ancestors: A Search for Who We Are
The Demon-Haunted World: Science as a Candle in the Dark

Acknowledgments

I hope it's clear that this book is a tribute and a love letter to my parents, Ann Druyan and Carl Sagan. I cannot begin to express my gratitude that in all the randomness I ended up with them. Nothing in my life would be possible without them. I am who I am because of their love, their wisdom, their generosity, and their confidence in me. My mother read several drafts of this book and helped me improve at every turn. In the years since my father's death, she has regularly left me in awe of her strength, creativity, and brilliance. She has worked tirelessly to carry on my father's legacy, but please know that it is her legacy just as much as it is his. She is a genius science communicator in her own right, as well as being the most enthusiastic cheerleader a daughter could ever hope for.

To the rest of my beloved family, especially Sam Sagan, Nick Sagan, and Clinnette Minnis Sagan, thank you for love, advice, and countless hours of stimulating discussion.

To my wonderful mother-in-law, Laurie Anne Robinson, thank you for all your kindness and help throughout this process, from often watching Helena to helping me track down some family history. To Laurie, as well as to my father-in-law, Andy Noel, and to Jon's grandparents, Margaret and Dwight Robinson, thank you for always making me feel welcome and for permission to share some of the details of your lives in this book.

To my talented editors, Tara Singh Carlson and Helen Richard, thank you for your ingenious revisions and thoughtful insights. This book would be vastly inferior in every way without your contributions. Thank you, as well, to my meticulous copyeditor, Kathleen Go. And to Kerri Kolen, who acquired my book for Penguin Putnam, I am eternally grateful.

My literary agents, Claudia Ballard and Eve Attermann, are encouraging and brilliant and I'm very lucky to work with them. And to Jennifer Rudolph Walsh, for helping me conceive the very first iteration of this book during one short conversation five years ago. Thank you to all three of them for making this book possible.

Many dear friends, gifted colleagues, and lovely strangers have been kind enough to impart some combination of their expertise, experiences, inspiration, insight, and support in the service of this book. They include (but are not limited to): Micah T. Adler, Katie Arnold-Ratliff, Liat Baruch, Mallory Bay, Brittany Berman, Zachary K. Blumkin, Mena Boyadzhiev, Lucy Boyle, Eric Brown, Aaron Chandler, Jessie Chasan-Taber, Chhobi Choudhury, Rita Choudhury, Leith Clark, Marlene Cohn, Loretta Deaver, Joy Fehily, Becca Fredane, Emily Fitzgerald, Brandon Kyle Goodman, Ben Grannan, Jedidiah Jenkins, Bahira Jawad, Luma Jawad, Mohammad Jawad, Daniel Ladinsky, Mimi Maritz, Alli Maxwell, Megan Perlman, Harry Petrushkin, Phoebe Segal, Rabbi Peter Schweitzer, Wenxiao Guo Tiano, Cathy Trentalancia, Tim Whitmarsh, Isabel Wilkinson, and Sara Zandieh. And to Richard Anderson, my favorite English teacher, from whom I learned so much, who died just as I finished writing this book.

Special thanks to Svetlana Zill, Safa Samiezadé-Yazd, and Andre Bormanis for their careful reading of the manuscript, and their help correcting my many mistakes as well as adding valuable information. And to Gynn Booth and Philippine Lelau, who made this book possible by lovingly caring for Helena during much of the writing of this book.

To my lifelong best girlfriends, Clara Hatcher Baruth, Amy Rosoff Davis, Lindsey J. W. Diamond, and Jessica Eth Jacobson, who convinced me on a daily basis that I could actually write a book, thank you. You'll never know how much your encouragement and collective comic genius helped me.

And to my husband, Jon, the love of my life, I do not know how I ever got so lucky. Thank you for your unconditional love at every turn, for the countless words of encouragement, for your endless patience with me, your confidence in me, and for making every single day worthy of celebration. As long as I get to spend it with you, one blink of an eye in the vastness is enough for me.

I'm deeply grateful to the following publications for permission to adapt essays I previously published there for this book:

Material adapted from the article "Love Song" by Sasha Sagan, originally published by *O, The Oprah Magazine*.
Material adapted from the article "The Empty Space" by Sasha Sagan, originally published by *O, The Oprah Magazine*.
Material adapted from the article "The Ladies Dining Society" by Sasha Sagan, originally published by *The Violet Book*.
Material adapted from the article "Instead of Christmas" by Sasha Sagan, originally published by *Literary Hub*.

ACKNOWLEDGMENTS

Material adapted from the article "Lessons of Immortality
    and Mortality from My Father, Carl Sagan" by Sasha
    Sagan, originally published online by *The Cut*, which is
    owned and operated by New York Media, LLC.
Material adapted from the article "Astrology vs.
    Astronomy" by Sasha Sagan, originally published by
    Wilderness magazine.